Routledge Revival.

Transportation and Urban Land

Urban land is a precious resource and originally published in 1961, *Transportation and Urban Land* aims to create an approach to analysing and projecting its uses with a particular focus on the household sector. By considering matters such as employment centres, organisation and technology of transportation and marginal valuation of residential space, Wingo develops a model to estimate how much land is required for residential land uses. This title will be of interest to students of Environmental Studies and professionals.

Transportation and Urban Land

Lowdon Wingo, Jr.

RFF PRESS
RESOURCES FOR THE FUTURE

First published in 1961
by Resources for the Future, Inc.

This edition first published in 2016 by Routledge
2 Park Square, Milton Park, Abingdon, Oxon, OX14 4RN
and by Routledge
711 Third Avenue, New York, NY 10017

Routledge is an imprint of the Taylor & Francis Group, an informa business

© 1961 Resources for the Future, Inc.

Publisher's Note
The publisher has gone to great lengths to ensure the quality of this reprint but points out that some imperfections in the original copies may be apparent.

Disclaimer
The publisher has made every effort to trace copyright holders and welcomes correspondence from those they have been unable to contact.

A Library of Congress record exists under LC control number: 77086416

ISBN 13: 978-1-138-96267-5 (hbk)
ISBN 13: 978-1-315-65920-6 (ebk)
ISBN 13: 978-1-138-96277-4 (pbk)

TRANSPORTATION

and

URBAN LAND

By LOWDON WINGO, JR.

RESOURCES FOR THE FUTURE, INC.

© 1961 by Resources for the Future, Inc., Washington, D.C.
Library of Congress Catalog Card Number 61-13662
Price $2.00

Second Printing, 1964

RESOURCES FOR THE FUTURE, INC.

1775 Massachusetts Avenue, N. W., Washington 6, D. C.

Resources for the Future is a nonprofit corporation for research and education in the development, conservation, and use of natural resources. It was established in 1952 with the co-operation of The Ford Foundation and its activities since then have been financed by grants from that Foundation. Part of the work of Resources for the Future is carried out by its resident staff, part supported by grants to universities and other nonprofit organizations. Unless otherwise stated, interpretations and conclusions in RFF publications are those of the authors; the organization takes responsibility for the selection of significant subjects for study, the competence of the researchers, and their freedom of inquiry.

This book is one of RFF's regional studies, which are directed by Harvey S. Perloff. Lowdon Wingo, Jr., is research associate in the field of regional economics.

Staff Editors, Henry Jarrett and Vera W. Dodds

Foreword

As more and more of the nation's activities and facilities center in our urban agglomerations, urban land increasingly must be viewed as a precious and scarce resource whose use and abuse are of the greatest public interest. The critical issue is how land is to be allocated and managed so that private equities are not abridged while the more general public interest is served. There is an urgency about this because of the extreme interdependence that characterizes urban activities, an inter-dependence that is dramatically illustrated by problems such as those of smog, traffic congestion, water pollution, and spreading blight.

An increasing number of decisions, private as well as public, are coming to depend on estimates of how various activities and facilities within the metropolitan community are likely to be arranged in space, and these estimates, in turn, depend on an understanding of the processes of urban location and arrangement. In the public sector such estimates are at the heart of city planning and development. Success of govern-mental efforts at guiding and controlling land uses depends in no small degree on how well changes in the patterns of settlement and activities have been anticipated; the design of land use policy depends on the local government's estimate of what is likely to happen in the absence of such policies, as well as on the community objectives. The soundness of the location of public facilities also depends on estimates of the growth pattern of the metropolis. The recent developments in the planning of urban transportation systems have greatly intensified the need to under-stand the processes of urban arrangement: the basic specification of the costly urban expressway networks is the spatial distribution of trans-portation demand, which, of course, depends on the future distribution of urban activities.

A host of troublesome questions have arisen in connection with the spreading out of urban developments into the rural hinterland. The growing literature on urban "scatteration" is a measure of the concern about a more "efficient" spatial organization of the city. A better under-

standing of how spatial patterns come into being and change is essential
to an operational formulation of what "efficiency" in this sense means
and how it relates to different forms of urban organization.

In the private sector, the changing spatial patterns of urban activities
and of land uses provide a framework for much that is done and not
done by individuals and firms. This is true of the choice of a place to
live, the choice of a retail site promising future advantages of market
access, or the choice of a plant location in the face of changes in the
distribution of the labor force or in the transportation system.

While none of these issues of urban organization is especially new,
the level of public awareness and the sensitivity of private decisions to
these urban processes have increased and the scale of the problem has
changed as well, particularly with the establishment of the national
highway program. This makes imperative a firmer understanding of
the processes involved and a better framework within which analysis
and policy appraisal may be carried out. The design of such a framework is a truly difficult task. The urban pattern is the result of a complex array of physical, economic, behavorial, and technological conditions interacting with each other. The key elements must be dealt with
if a basis for making estimates (which, as suggested, is at the heart
of so many urban decisions) is to be provided.

The present study represents an attempt to develop such a framework
for analyzing and projecting urban arrangement. Lowdon Wingo has
made an original and significant contribution. An earlier draft, distributed to scholars last year, generated interest and discussion, so that
the study promises to open up a valuable line of productive work.

Because of the great complexity of the problem, the author has chosen
to illustrate an approach rather than attempt to detail a complete
framework. He has focused attention on the largest use of urban land,
that for residential purposes by households. This use takes up, on the
average, some 40 per cent of the total developed land and roughly 80
per cent of the land area in all private uses in urban communities. A
significant portion of all the other uses, such as for streets and shopping
centers, is closely related to dwelling use. If we can increase our predictive skills in defining the nature of the demand for urban space by
the household sector, it will improve our ability to predict the land
requirements and patterns which emerge from urban growth and change.

He identifies certain critical considerations in arriving at the amount,
distribution, and value of land required for residential uses, namely,

(1) the spatial pattern of employment centers, (2) the organization and technology of transportation, (3) the marginal valuation of residential space by the household, and (4) the marginal valuation of leisure by the worker. While the model he develops relies on basically simple and somewhat abstract concepts, and while certain important factors in the locational decisions of households—such as prestige, culture group associations, and variations in the quality of highly valued local services—are left out, the approach he develops clearly lends itself to enrichment and expansion.

The great value of the Wingo model is that he brings to the forefront critical elements—such as the relative value placed by persons on time and on space around the home, and the technology of transportation—which, while generally acknowledged as important, tend to be left out in arriving at land use and transportation policy decisions. He brings such elements to the center of the stage by showing how they are related to considerations which are already a focus of attention, such as the cost of land and the relative pulling power of the central business district. By employing a partial equilibrium model, he can show how technological and behavioral factors influence, and are influenced by, changes in relative prices and, in general, developments within the urban economy.

The model lends itself quite readily to the inclusion of additional considerations; it is basically flexible and open ended. It is our hope that other scholars will be challenged to expand the model and, of course, to improve on the approach and method itself and that empirical testing of this and (hopefully) more advanced models will be tried out by operating agencies as well as individual scholars.

Among other things, the present study will delight all those who appreciate a sharp and logical "solution" of a tough, complicated, and extremely significant problem.

HARVEY S. PERLOFF

Preface

This monograph is the product of a manifold interest at Resources for the Future in problems of the metropolitan community. First, the use, regulation, and conservation of urban land involve important national resources with which the organization has an overriding concern. More specifically, the Committee on Urban Economics of Resources for the Future has a special interest in problems related to intra-metropolitan organization and change. In addition, the underlying interest of the author and many of his associates at Resources for the Future in problems of planning and policy-making for the metropolitan community has provided a strong motivation for taking a new cut at a bundle of old problems.

Gratitude is owed to many for this final product. Those who took charge of the editing and production of the monograph deserve special commendations. In terms of substance, the advice of many persons was of inestimable value. Among these were my associates on the staff of Resources for the Future, many of whom gave freely of their time to discuss critical points in the thesis. Several scholars reviewed an earlier draft, and their detailed comments set the guidelines for this final version. Others permitted me to presume upon friendship or professional association to exploit their special competences. Of all these, I am most deeply indebted to Harvey S. Perloff, my friend and colleague, for without his warm encouragement this monograph might never have been written.

LOWDON WINGO, JR.

Contents

Transportation and Urban Land

I

Introduction

Urban Planning in a Changing Policy Environment

The enactment of the National Interstate and Defense Highways Act of 1955 brought to a close an era in urban planning and land use policy, and its implementation in the following years set off a revolution in our thinking about the urban environment which is just beginning to coalesce into a new intellectual framework. Previously the effects of any single decision, public or private, on the total pattern of metropolitan growth were likely to be relatively small and localized, and the pace at which they worked themselves out almost leisurely. Plans were built up by a succession of small decisions, taking off from the massive "given" of the status quo and frequently looking much more to the parts than to the whole, if only because the process of design with its emphasis on a visual aesthetic made much more sense at the level of the neighborhood than at the expanded level of the great metropolitan complex. Basic problems for city planning were how to provide for an orderly occupance of the hinterland, how best to "renew" the obsolescent parts of the city, how to provide effective coordination of the plans of local agencies so that over a period of time they would add up to an urban environment not inhospitable to its citizens.

The Highway Act changed all that by releasing the tremendous forces for urban reorganization previously held in check by the restraining bonds of existing opportunities for transportation and communication. It soon became apparent that a single great urban expressway could as ruthlessly demolish a carefully developed "master plan" as gouge its way through the countryside and the cityscape. The scope of the forces released and the rapidity with which their effects developed did not allow for a thoughtful and measured adjustment, and in many cases new

1

plans and policies were thrown together in weeks rather than years, based on hasty research into problems which even now are not well understood.

Consider some of the effects of the National Highway Program on urban planning. In the first place, justification for the urban phases of the national program rests on the primacy of the national interest in efficient urban transportation, and interest shared by the states as the executors of the program. Since the power of decision rests with the nation, the degree to which local interests are considered depends on the vagaries of intergovernmental relations, so that a major part of the local planning effort must center on the adjustment of plans and policies to the consequences of decisions from the outside. Then there are the vast, if invisible, changes which new facilities have induced in the processes of urban organization—the new prime land made accessible for development, the older areas more acutely disadvantaged, the competitive opportunities redistributed. Old trends become discontinuous, and in the period of uncertainty before new trends manifest themselves planning is most difficult, and most imperative. Finally, as new decisions have altered the urban processes, so have they vested new interests and dispossessed old ones. Newly advantaged groups see these new forces as progressive, and the planner—the proponent of order and balance—as the stubborn and reactionary opponent of progress. For the planner, then, the issue is not where to locate a transportation facility but how to adapt to the new policy environment which the National Highway Program has brought into being.

Perhaps the most significant aspect of this development is the change in intellectual demands posed by the new policy issues: increasingly they are issues which are not easily amenable to conventional design solutions but require new kinds of knowledge, new frameworks of analysis, and new criteria for making decisions. Improved understanding of how parts of the urban society or economy react to changes in the important dimensions of the urban environment is required: how the urban land market reacts to changes in the accessibility of land; how changes in transportation are likely to affect a firm's choice of location; what the recreational needs of urban groups are; how long-run secular national and regional trends affect the urban environment. Knowledge in these fields is increasingly needed in fashioning intelligible urban policy, especially in planning public facilities and in regulating land use.

Equally important are the changes that are emerging in the analytical and policy frameworks of planning. Where once the process involved a comparatively simple framework of defining objectives, identifying the relevant features of the status quo, designing a plan to relate them, and from this drawing the relevant policy recommendations, now the planner is confronted with mathematical models, computers, theories from alien disciplines, all of which were ushered in with the accession of the highway engineer to a dominant position in urban decision-making. But these were really changes in technique only—the highway engineer used basically the old planning framework, defined his narrower objectives in a mathematically more sophisticated manner, engineered a solution which was fundamentally a design solution, and then recommended the investment of public funds to carry out the design. The real revolution began when he began, responding naturally to the interrelatedness of urban means of transportation, to design transportation systems which encompassed the entire metropolis and involved gigantic programs of public investment. This step had two important effects: (1) the objectives of the highway engineer, and the programs by which he proposed to achieve them, soon clashed with the cherished goals not only of the city planner but of important interest groups in the community; and (2) the planning of transportation systems evolved rapidly from the laying out of freeway nets to the planning of region-wide, integrated, multi-modal systems which required kinds of decisions which the engineer was not qualified to make—decisions whose secondary effects on the urban pattern were much more important than the problems they sought to remedy.[1] It is likely that in the end these two developments will reinvest the urban planner with his policy role, but under circumstances requiring him to work in a vastly more complex framework. The old trinity of "objectives, status quo, and plan" will not suffice to provide rational policies in an urban universe of interrelated private and public

---

[1] ". . . it is intended not only that the recommended transportation system should provide convenience and economy of travel, but also that its influence on the development of the area should tend toward facilitating a *desired pattern of regional development."* (Italics mine.) Penn-Jersey Transportation Study, *Prospectus* (Philadelphia, 1959), p. 2.

"Technically, the point has been reached where the entire circulation of the community can be comprehended. Whole systems of streets and transit facilities, of spatial locations of activities, and of interchange of persons and vehicles can be dealt with. Careful planning with this enlarged scope offers greater certainty of acting in the public interest than the method of making piecemeal adjustments in response to local pressures." Chicago Area Transportation Study, Volume I, *Survey Findings* (Chicago, 1959), p. 93.

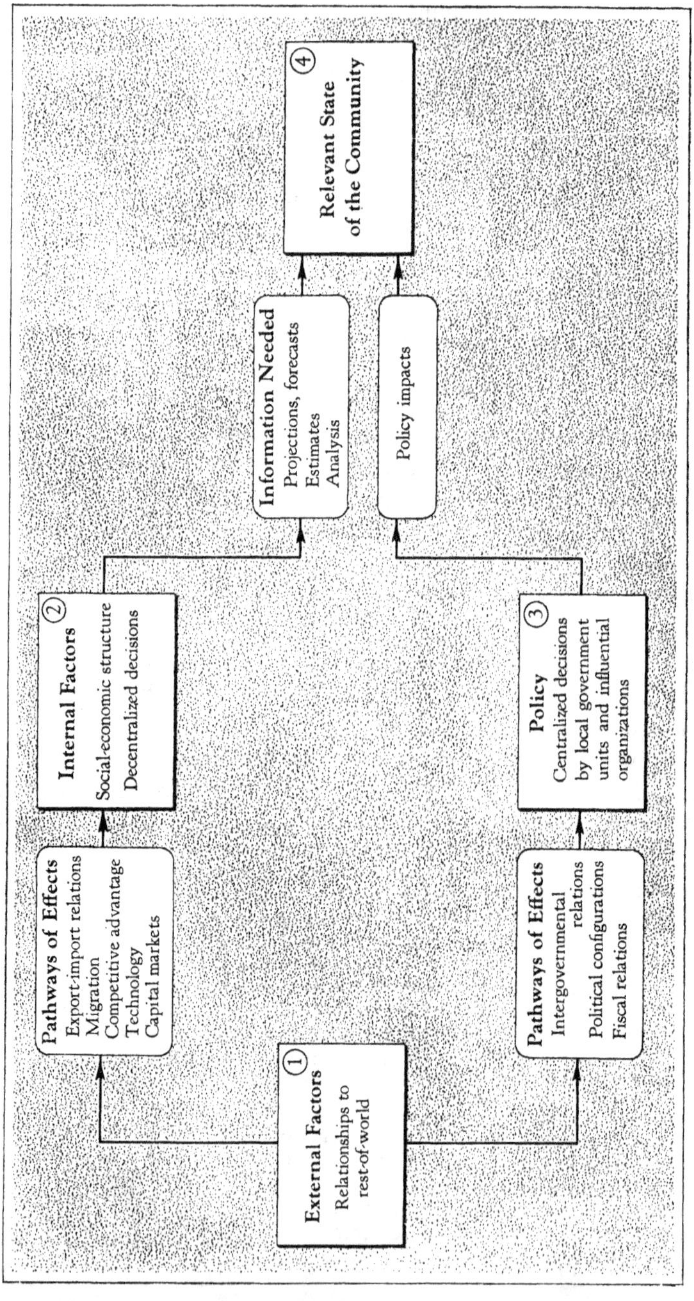
1
External Factors
Relationships to rest-of-world
Pathways of Effects
Export-import relations
Migration
Competitive advantage
Technology
Capital markets
Pathways of Effects
Intergovernmental relations
Political configurations
Fiscal relations
2
Internal Factors
Social-economic structure
Decentralized decisions
3
Policy
Centralized decisions by local government units and influential organizations
Information Needed
Projections, forecasts
Estimates
Analysis
Policy impacts
4
Relevant State of the Community

FIGURE 1.

purposes and of intricately related developmental processes which can be irreparably disturbed by the effects of poorly calculated "big" decisions.[2]

The New Policy Framework

Enough of the dimensions of a new framework have emerged so that we can talk about its formal characteristics with some confidence. Assume that we have four components, or boxes, wired together and labeled as follows (see Figure 1): Box 1 is called "External (or rest-of-world) Factors" and is wired directly to Box 2, "Internal Factors" (such as the local economy, local social structure and its processes, or any part of these) and to Box 3, which we will label simply "Policy." Both Box 2 and Box 3 are wired directly to Box 4, "Relevant State of the Community." Box 4 really expresses the *joint* consequences flowing from Boxes 2 and 3, each modified by the flows from Box 1. It is possible, of course, to talk about all sorts of complex feedbacks and side effects, but our concern is with the processes of urban planning as they are carried out in this kind of framework. To carry the analogy a step further, assume that effects constantly flow over these circuits and affect the readings of a series of dials on Box 4, while on Box 3 there are a series of switches which will modify the flow from 3 to 4. The process of planning is analogous, then, to determining what readings we want to achieve on the Box 4 dials, anticipating the flows 1–2, 2–4. and 1–3, and setting up a schedule for manipulating the switches on Box 3 that will achieve the target readings on Box 4's dials.

This is, of course, a highly simplified analogy of the emerging framework for planning, but it shows that the new framework is substantially different from the old in several critical respects. First, goals are limited to those readings which are possible on the dials of Box 4: the planner is constrained constantly to formulate objectives which the pressing realities of society, of technology, and of resources permit.

---

[2] "Past a certain point, however, the paring down of transportation facilities will inhibit population growth and employment opportunities and perhaps impair the average productivity of the region's workers. Certainly, it may absorb an undue amount of their time in the journey-to-work." *Preliminary Financial and Organizational Report Regarding Metropolitan Transportation*, 86th Congress, First Session, Joint Committee on Washington Metropolitan Problems, August 1959 (prepared for the Committee by the Institute of Public Administration, New York), p. 3.

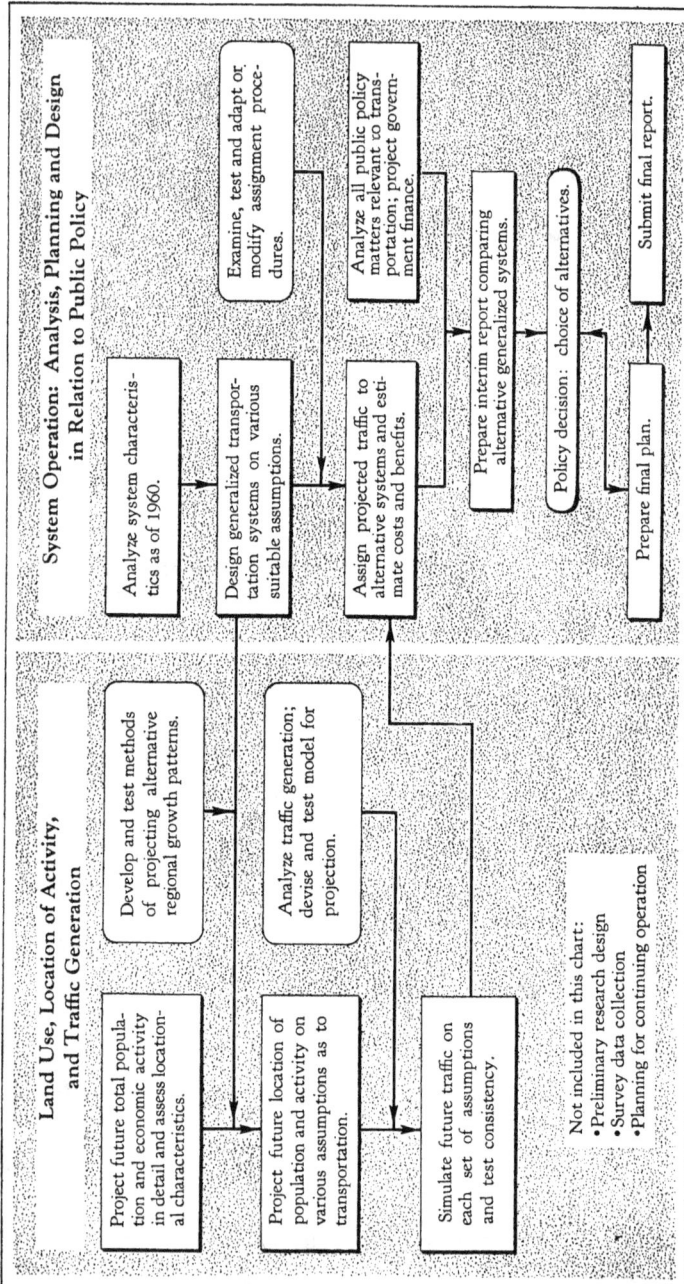

FIGURE 2. *Condensed flow chart on analysis and planning program for a transportation system.*
Source: Penn-Jersey Transportation Study, *Prospectus* (Philadelphia, 1959), Appendix C.

Second, he is confined to the number and nature of the switches on Box 3, that is to say he is constantly bound by constitutional, statutory, and institutional limitations as well as by the availability of fiscal resources. Third, the readings on Box 4 are the *joint* result of the flows of effects from Boxes 2 and 3, rather than the simple consequences of throwing switches on Box 3. In short, plan objectives are not achieved by the effectuation of planning policies alone but must be posited also on autonomous social and economic developments in the community at large. Finally, one must always calculate the flows from Box 1 to Boxes 2 and 3: both the objectives and the plan must be intimately related to the powerful trends and the dramatic new developments in the regional and national environments.

At one level, of course, this expression of a new framework really doesn't tell the planner anything he didn't learn in school; one must be conscious of all the salient developments that affect the form and processes of urban growth. At another level it is a useful innovation, for it focuses attention on the interrelatedness of public and private decisions and on the critical linkages of the urban community with the world at large. In short, this is a highly generalized decision-making model,[3] but more, one which invites the application of powerful quantitative methods of analysis as well as the extensive integration of broad policy sectors.

The Penn-Jersey Transportation Study's *Prospectus* clearly demonstrates the impact of the new developments, as Figure 2 shows. Here it is proposed to project the level, composition, and locational characteristics of the local economy against those of the national and regional economies (Flow 1–2, as shown in Figure 1), to develop and try out a number of transportation policy alternatives (switching patterns in Box 3), and to forecast the distribution of the local economy's activities within the metropolitan area (Box 2); the generalized consequences of these steps (readings on Box 4) will be the basis for choice for a policy committee.

In addition to these changes in the framework for planning, changes are taking place in the very bases for decision themselves. One does not merely make decisions which effectuate policies to achieve ends. Policies have broad streams of consequences, and, as our experience in

[3] Werner Hirsch is entitled to credit for the first statement of this generalized, analytical-decisional model in a paper, "Regional Interaction Analysis," presented to the Committee on Regional Accounts, Resources for the Future, Inc., Spring 1959.

transportation planning has proved, one cannot exclude from consideration all effects except those which achieve, or are likely to achieve, the
ends set forth. Increasingly the planner will become preoccupied with
"side effects," indeed with the whole stream of policy consequences, and
he will have to develop new ways of choosing among alternative policies.

Some promising advances in rationalizing urban planning may be
anticipated from the extension of cost-benefit analysis. But promising
as cost-benefit analysis is, it does not by itself solve the whole problem.
It provides useful criteria for investment where direct effects of the
investment can be identified, important side effects are simple and easily
quantifiable, and the investment is planned for the near future. It seems
certain that its usefulness will be increased by the extension of the range
of benefits comprehended and improvement of techniques of valuing
them. Such investment activities are, however, only part of the planner's
policy concerns, the other major part being his responsibility for land
use and similar forms of regulation. Costs and benefits are certainly
relevant to urban planning decisions, but only in a minor way as investment criteria. The problem is to identify the redistributive effects of
regulation and to establish criteria of equity in the policy redistribution
of values.

The simple fact is that the new framework will of necessity require
new decision criteria. Not only will we need to know how to choose
among the great and gross alternatives of metropolitan form, but we
must develop much more sensitive techniques for choosing between the
complex streams of consequences that flow from alternative policies.
To "accord with the master plan" is no longer sufficient justification for
policy, for one of the subsidiary results of the changing intellectual
environment is an awareness that the planner must become *more responsible* to the total community for the total consequences of his plans; a
freeway gouging the heart out of a viable residential neighborhood requires more than coherence with a master plan to justify it. The new
responsibility certainly requires that the human costs to minorities
"redeveloped out" of urban centers in favor of upper income groups be
calculated as part of the decision to redevelop.

New kinds of knowledge, new frameworks of analysis, new criteria
for decisions, new responsibilities—these are urban planning's inheri-
tance from the National Highway Program. In general, a new logic is
required to permit the planner to exploit the rapidly growing array of
powerful quantitative techniques which can extend the range of his

thought and amplify his effectiveness as a critical policy influence on the urban scene. These considerations set the stage for this study. The purpose of the study is to treat a major problem for planners—the location of economic activities within the metropolitan area—the framework of a quantitative model. More specifically, the model will be designed to analyze how the household sector of the urban economy arranges itself in space, to identify the critical variables, and to suggest the consequences flowing from policy alternatives.

The strategy for this study is to articulate some of the "relationships that count" into a simplified view of the problem consistent with theory in the social sciences. The framework will comprehend these relationships as a system in which quantitative changes in one sector will result in changes in spatial arrangements which can be quantitatively described. An appropriately designed mathematical model can provide such a framework. How it can do this calls for a brief digression on some relevant characteristics of economic models.

A Note on Models

A model is simply a way to express significant causal or structural relationships stripped of the irrelevancies and complexities of the real world so that they may be more readily understood:

> What we have a right to ask of a conceptual model is that it seize on the strategic relationships that control the phenomenon, and that it thereby permit us to manipulate, i.e., think about, the situation.[4]

A mathematical model, more specifically, is a set of quantitative relationships expressed in the language of mathematics and describing the interaction of phenomena—in this case of social phenomena.[5] It consists of operations, variables, and parameters. The operations identify the nature of the relationships (whether variables are to be added, multiplied, raised to specified powers, and the like). The independent variables are those whose values are determined outside of the model and "fed into" it as data inputs; dependent variables are those whose values are produced by the model as a result of the data inputs. Finally, the

[4] Robert Dorfman, Paul A. Samuelson, Robert M. Solow, *Linear Programming and Economic Analysis,* The RAND Series (New York: McGraw-Hill, 1958), p. 9.
[5] Cf. Kenneth J. Arrow, "Mathematical Models in the Social Sciences," in Daniel Lerner and Harold J. Lasswell (eds.), *The Policy Sciences* (Palo Alto: Stanford University, 1951).

parameters are the "weights" of the operations, which represent the quantitative expression of the relatively fixed conditions within which the model operates; they are, in effect, arbitrary values assigned to the operations to make the model correspond with observed behavior.

To illustrate these elements, take a simple model such as one recently suggested by Donald Bogue:[6]

$$L = kP$$

where L = acres of land in farms converted to nonagricultural use,

P = increase in total population of the SMA, and

k = an empirically determined constant.

This model says that the amount of farm land converted to nonagricultural use in any small metropolitan area (dependent variable) is equivalent in volume to the increase in population times some constant "k" (parameter). From census data Bogue determined that "k" had a value between .172 and .265 with a median value of .238.

The distinction between parameters and variables is especially pertinent to our problem and so merits additional attention. In the model above (and for the circumstances defined) the value of "k" remains unchanged; in different circumstances, however, it might have quite different values. In one sense, parameters are merely less variable variables. Where parameters reflect the structural relationships between variables, the parameters will change as the relationships change. Hence, parameters change slowly compared to variables and so may have approximately constant values in the short run. As the time period is extended, autonomous structural changes may accumulate in the underlying phenomena, resulting in increasing divergence of parameters from the conditions originally specified.

But there is another way in which the circumstances surrounding the variables may be changed: it may be possible to intervene in them and modify them so considerably that the values of the parameters are substantially altered. Where such intervention can be effected with some precision with regard to the parameters, they become, in effect, "policy variables." So, for example, if land were nationalized in the United States, the value of "k" in the above model would reflect national policies concerning land utilization and not the end result of certain complex

---

[6] Donald J. Bogue, *Metropolitan Growth and the Conversion of Land to Non-agricultural Use* (Scripps Foundation Research in Population Problems, Miami University, and Population Research and Training Center, University of Chicago, 1956).

market phenomena governing the conversion of land in the metropolitan periphery.

Hence, the variables taken by themselves are of interest and use essentially in the short-run predictive and analytical sense. Long-range forecasting and public policy share a common interest in "parametric" changes in the model, the former in terms of the autonomous structural changes that are working themselves out in interrelationships among the variables, and the latter in the sensitivity of certain of the governing circumstances to deliberate social action to change them. Consequently, in the design of the model it will be important to distinguish between the structural and policy conditions which are reflected in the parameters. In this distinction rests our ability to describe how market conditions and public policy interact to develop the spatial arrangements in which we are interested.[7]

In short, the model set forth in this study will take certain demographic and economic variables as independent. The pertinent assumptions will be expressed by three kinds of parameters reflecting policy, market, and technical conditions. The output will be a set of values expressing for every point in the urban space a unit value and a density. Before we proceed with the actual model, however, a brief look at the present state of analysis will provide a useful perspective.

Some Alternative Approaches

Several approaches have emerged recently for estimating the future distribution of households in the urban region, three of which warrant some attention: *ad hoc* allocations; gravity, or potential, models; and economic models. Most of these formulations identify the problem as one of distributing a given number of *new* units (new households, dwelling units, or population) among a number of small areas into which the urban area has been divided. It is important to note that effective techniques to handle in systematic fashion parametric changes—i.e. changes tending to redistribute activities or households—have not been developed. The volume and direction of these changes are difficult to detect at any moment, and the subtlety of factors leading to them, such

---

[7] Although market conditions have been identified with structural parameters, another kind of structural parameter will command our attention, namely, technical parameters describing relevant features of technology and spatial relationships. These might be thought of as the engineering conditions of the model.

as alterations in technology, in the organization of production and distribution, and in the behavioral characteristics of the community, makes their anticipation a formidable problem.

AD HOC ALLOCATIONS

In urban land use planning, "best practice" in the spatial allocation of new dwelling units has been described by F. Stuart Chapin, Jr.[8] It involves two general steps: (1) the determination of an effective net "holding capacity"[9] for each of the subareas, or planning districts, in accordance with the amount and quality of available vacant and renewable land, prevailing densities, and zoning and density standards; and (2) allocation of the estimated total of new households among housing and density types in accordance with density standards, local trends and conditions, and population-income factors, and then among planning districts in accordance with existing building trends, access conditions, and service availability, as well as other relevant conditions. This is not so much a model as a system of successively disciplined judgments, intuitive as well as objective, about how the city is likely to grow—at each stage a particular kind of assumption is made. The merit of the technique is that it identifies the important assumptions and requires them to be explicitly stated.

More recently, a modified technique was evolved by the Chicago Area Transportation Study for the purpose of predicting the Area's 1980 traffic requirements.[10] Like the Chapin technique, it requires that a holding capacity be defined for a large number of small areas. The distance of the small areas from the central business district is then related to the percentage of capacity its present population represents (Figure 3). For a target year (1980) this function is adjusted by estimation of the distance and per cent coordinates of the 1980 peak and the nature of the slope in both directions from the peak so that the adjusted curve represents the estimated target year population distributed among concentric zones (Figure 4). Each small area is then assigned additional

[8] F. Stuart Chapin, Jr., *Urban Land Use Planning* (New York: Harper & Bros., 1957), pp. 339-54.

[9] "The number of dwelling units the vacant and renewal land . . . will accommodate according to a prescribed pattern of residential densities." *Ibid.,* p. 350.

[10] John R. Hamburg and Roger L. Creighton, "Predicting Chicago's Land Use Pattern," *Journal of the American Institute of Planners,* May 1959, Vol. XXV, No. 2, pp. 67-72.

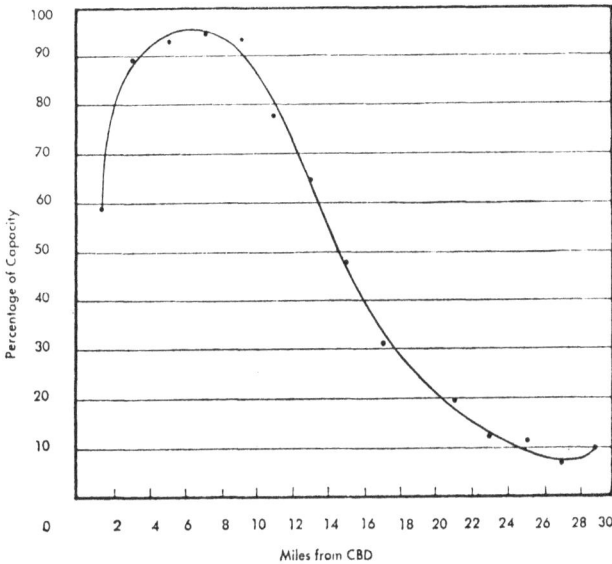

FIGURE 3. *1956 population as a percentage of population capacity (by distance from the central business district).*
Source: John R. Hamburg and Roger L. Creighton, "Predicting Chicago's Land Use Pattern," *Journal of the American Institute of Planners,* May 1959, Vol. XXV, No. 2, pp. 67-72.

population up to the point that its population level is that proportion of its holding capacity indicated by the developed target year function above.

The Chicago Area Transportation Study technique relies less on arbitrary judgment and more upon a hypothecated process of growth than does the land use planning approach and so more closely approximates a predictive model. It has not completely excluded subjective judgment but confines it largely to the design of the concentric "per cent of holding capacity" function for the target year. In short, given current population and land use data, one needs only to feed into this framework the target year total population estimate and to design this arbitrary function, and this mechanism will provide a distribution of population (households) throughout the urban area. This technique sees the urban growth process essentially as a great expansion of densities in a con-

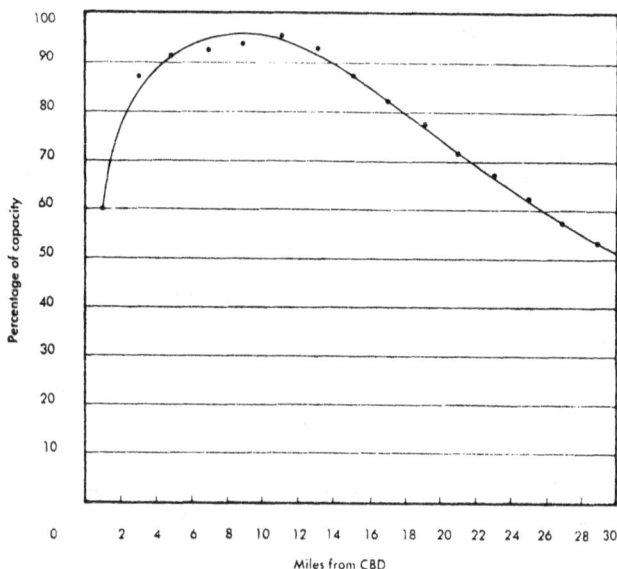

FIGURE 4. *Estimated percentage of population capacity (by distance from the central business district, 1980).*
Source: Same as Figure 3.

centric pattern about an inner zone in which densities are estimated to decline.[11] As a model, the Chicago approach tells little about how the distributive processes shaping the urban pattern actually work. Indeed it offers little in the way of hypotheses for research. Its importance lies in its statement in mathematical language of a concept of how urban patterns change under conditions of growth.

THE POTENTIAL MODEL

The potential or gravity model postulates a declining relationship of a phenomenon to distance from a given focal point and has been applied to a considerable variety of social phenomena with an impressive degree

[11] "Both these distributions [population and manufacturing workers] illustrate the tremendous organizing impact of the central area on . . . the development of land. While there are sectoral effects present, these are assumed to be distortions of a purely concentric ring growth. . . ." *Ibid.*, p. 70.

of success.[12] Walter Hansen has applied the potential model to the problem of urban spatial organization.[13] Hansen's model states that if the urban area is cut up into a number of zones, for each of which employment, travel time from it to all other zones, and the amount of vacant, developable land can be determined, an estimated increase in total population can be distributed among the small areas on the basis of a positive relationship to "accessibility" and to the available vacant land.

The Hansen model has several important attributes compared to the Chicago model. First, it has greater behavioral content—how families choose their locations *is* related to access to employment and to the locational opportunities available in various parts of the city. Second, it has received some testing of its predictive power and hence we have some idea of its reliability.[14] Third, it acknowledges a more sophisticated set of variables and their relationships, such as the distributive effects of the transportation system and the spatial array of employment. Fourth, it is more neutral with respect to its output than the Chicago model, whose output is shaped by a preconceived view of what the output should look like. A final characteristic is worth mention: the output of the potential model is dependent on the "scale" of the input. One will get different results from playing through the model a population increase of 10,000 than by successively iterating it in two operations of 5,000 each. The choice of the "iteration pattern"[15] is therefore an act of judgment which will influence the results. This feature gives the model a dynamic quality not implicit in others: it allows successive distributions

[12] The definitive work is G. K. Zipf, *Human Behavior and the Principle of Least Effort* (Cambridge: Addison Wesley, 1949). For general discussion and bibliography see Gerald A. P. Carrothers, "An Historical Review of the Growing and Potential Concepts of Human Integration," *Journal of the American Institute of Planners,* Spring 1956, Vol. XXII, No. 2, pp. 94-102, and Walter Isard, *Location and Space Economy* (Cambridge: Technology Press, 1956), Chapter 3.

[13] Walter G. Hansen, "How Accessibility Shapes Land Use," *Journal of the American Institute of Planners,* May 1959, Vol. XXV, No. 2, pp. 73-91.

[14] "This method was used to estimate the residential growth from 1948 to 1955 for each of the zones in the Washington metropolitan area. A comparison of the estimated growth to the actual growths showed that 40% of the zonal estimates were within 30% of the actual growths and 70% of the zonal estimates were within 60% of the actual figures." *Ibid.,* p. 75.

[15] The term "iteration pattern" refers to the manner in which a data input is applied to the model, involving how many fractions of a major input are to be successively fed into the model and, if the fractions vary, what sequence is to be followed. The choice of iteration pattern is an important decision in incremental models, especially under conditions of parametric change.

to be influenced by changes accumulated from those preceding. In general, the Hansen potential model is the first integrated mathematical model designed to deal with the distribution of households on the urban landscape.

ECONOMIC MODELS

The models described above say very little about how the individual unit is likely to behave, asserting that the collective behavior of new units will be similar *in the aggregate* to that suggested by the model, even though the model and real world processes may be widely dissimilar. Thus, these models do not say much about the processes resulting in urban spatial patterns. The economic models introduce behavioral postulates of general economics into the problem: each unit (individual, firm, industry) chooses the "best" alternative for itself among the choices available to it. The spatial pattern of the urban economy is thus a consequence of the optimum-seeking activities of individuals, each bidding for its "best" combination of two variables: quantity of space, and location of that space vis-à-vis all other units. The allocation takes place through the market mechanism moving toward an equilibrium in which space is allocated among users so that no two users can be made better off by exchanging combinations.[16]

How much space a unit will use depends on its "internal economy"— for the firm, its technology and scale; for the household, its consumption preferences. To the extent that unit behavior in the use of space varies with such elements as relative price, the demand for space is a useful concept. Locational requirements, on the other hand, arise from the external economic relationships of the unit to other units and the simple fact that to move goods, people, and information through space is a costly process which varies with the distance involved. The firm will, *ceteris paribus,* seek to minimize its total transport costs by exploiting the least-cost location. For the houshold, also, transport costs can be viewed as offsets against total satisfactions, so that the household is similarly motivated to seek a least-cost position.

---

[16] "Who gets a location in the narrower sense, a site? *He who is prepared to pay the most for it.* This has two implications: First, with prevailing price relations the individual in question finds his highest utility here. Secondly, at the price paid for this site he alone finds his highest utility here." August Lösch, *The Economics of Location* (New Haven: Yale University Press, 1954), p. 254.

The spatial "behavior" of a unit is a matter of its relating its internal economy to the economic environment in the most profitable fashion, by seeking to purchase space and location to maximize its net returns, or satisfactions. The process is carried out through the market, which, hence, allocates land and location to the most efficient users. This, then, is the basic framework within which economic models of urban arrangement have been developed. As yet none of these is "operational," but a beginning has been made, and the general characteristics of these models are emerging. They have been expressed in two basic forms: (1) the classical equilibrium model, and (2) the linear programing model.

The first class is typified by William Alonso's economic analysis of the urban land market,[17] which takes the form of a mathematical model focusing on the interaction of urban land rents, intra-urban location processes, and the intensity of use (or density of occupancy) of urban space. A more general treatment emerges from Richard Muth's recent work,[18] which sets up a simplified equilibrium model to examine the process of conversion of land from rural to urban uses. His model follows the form of the agricultural land use model advanced by J. H. von Thünen[19] in that it generates concentric bands of land use about a central point on a homogeneous plain. Two industries, one producing an agricultural commodity and the other nonfarm housing, have productivities which decline with distance from the market point. Given technology, factor prices, and demand conditions, a structure of land rents is generated, a boundary line between the industries emerges, and the prices and volumes of output of each are determined. Muth's model places the problem of urban arrangement within the framework of economic theory as one dimension of the local economy and its ties to the rest of the world.

Recent developments in urban studies have focused attention on the construction of an operational linear programing model. Stevens and Coughlin have discussed a model to allocate increments of industrial

[17] William Alonso, *A Model of the Urban Land Market: Location and Densities of Dwellings and Businesses* (unpublished Ph.D. dissertation, University of Pennsylvania, 1960).
[18] Richard F. Muth, "Economic Change and Urban vs. Rural Land Conversions," *Econometrica*, January 1961, Vol. 29, No. 1, pp. 1-23.
[19] J. H. von Thünen, *Der Isolierte Staat in Beziehung auf Landwirtschaft und Nationalökonomie* (Hamburg, 1826).

18 TRANSPORTATION AND URBAN LAND

activity among a number of subareas in the metropolitan region.[20] It allocates to minimize the total number of services required to transport products among the plants and to distributors. Herbert and Stevens have suggested that the household sector might similarly be treated, taking the household's transportation costs and housing demands as the significant features to be observed.[21] Such a formulation would seek to maximize some function of location rents, space per household, and accessibility to employment, subject to restraints involving the "supply of space," budgets, and quotas setting certain minima. The major difficulty in applying the linear programing framework to the household demand problem is that demand behavior is fundamentally nonlinear in nature, and a linear approach can be applied only if important features of consumer behavior are defined out of the model.

Elements for a New Model

These approaches suggest the variety of frameworks emerging. So far, these have been of two kinds—predictive and analytical—only because the state of analysis has been comparatively primitive. The overriding objective ahead is to develop operational models consistent with the relevant sectors of behavioral theory, models which are reliable for prediction and fruitful for analysis, for enriching our understanding of how cities grow and change.

Indeed, we can enumerate specific objectives for further work. First, the distinction between policy and structural effects needs to be sharpened. This need becomes especially critical where the ability to intervene is changing, as in the recent history of land use regulations in the United States. As policy effects expand, the structural effects of the market will perform a declining role in allocation and organization.[22]

Second, the analytical content of these models needs to be expanded. Hopefully, in the future their performance will be improved by extension of their range into more subtle behavioral realms so that the important

---

[20] Benjamin H. Stevens and Robert F. Coughlin, "A Note on Inter-Areal Linear Programming for a Metropolitan Area," *Journal of Regional Science*, Spring 1959, Vol. 1, No. 2, pp. 75-83.

[21] John D. Herbert and Benjamin H. Stevens, "A Model for the Distribution of Residential Activity in Urban Areas," *Journal of Regional Science*, Fall 1960, Vol. 2, No. 2, pp. 21-36.

[22] Cf. Ralph Turvey, *The Economics of Real Property* (London: Allen and Unwin, 1957), pp. 125 ff.

effects of such phenomena as income distribution, minority group segregation, and family composition can be included. The immediate concern, however, is to expand the variables and relationships of strategic importance in the gross characteristics of the spatial pattern and to provide for a "smoother meshing" of the engineering and economic components.

Third, it would be useful for a spatial model to fit within the broader framework of the urban economy. Development of major metropolitan economic studies is accelerating this movement by recasting the problem as one dimension of the total urban economy.[23]

Finally, the virtuosity and reliability of the models may be improved by application to space distribution problems of new techniques of analysis, such as the development of "simulation models" seeking to repair a basic flaw in single-valued, deterministic models—the "unreal" assumption of rational, homogeneous behavior of all units.[24]

The model to be developed here will seek to move in the first three of these directions: to provide a more explicit differentiation of policy and structural effects; to enhance its analytical value by bringing the main elements of the problems within the framework of economic theory; and in doing this to set up the conditions for treating the problem of intraurban distribution of population as a part of the urban economy as a whole. It will apply no new or radical tools of analysis, leaving to the sophisticated and perceptive reader to note where such techniques would provide more versatile ways of dealing with features of the model.

The construction of the model will proceed through development of components to describe the crucial relationships among the variables—relationships which are rooted in four general aspects of the urban environment. At one extreme are the physical characteristics of urban

[23] "To a large extent [the question of area distribution of economic activities within the Region] can be regarded as dependent upon . . . the analysis and projections of the aggregate Regional Economy: i.e., concerned with allocating to subareas of the Region various independently-determined Regional aggregates of activities. Certain feedbacks from the intra-Regional to the aggregate analysis are, however, essential. . . . For example, the availability and cost of large industrial sites depends in part on land use patterns in the Region and might be a significant factor influencing aggregate industrial growth." Pittsburgh Regional Planning Association, *Design for Economic Study of the Pittsburgh Metropolitan Area* (Preliminary), November 1959, p. IV-B-1.

[24] Britton Harris has an excellent brief discussion of the stochastic characteristics of the problem in the previously cited "Summary . . . ," and the actual construction of such a model has been proposed by William L. Garrison in his "Notes on the Simulation of Urban Growth and Development," Discussion Paper No. 34, Department of Geography, University of Washington, February 1960.

space, at the other a population of specified composition, and relating them are the relevant features of technology to overcome the frictions of space, and of the organization of society to produce goods and services. The spatial-technological end of this array yields engineering relationships that will interest us, while the economic organization of urban society is the source of the economic relationships. The problem is to translate a sector of the urban economy into spatial dimensions through an appraisal of the strategic relationships imposed by technology.

Physical space enters the model in two forms: as a resource, a divisible commodity with simple quantitative characteristics;[25] and as a locational matrix in which its relational qualities are critical. "How much" and "where" are each spatial problems which the model must synthesize.

Economists have rarely been at ease with problems involving spatial relationships, and even the theorists who have made the most useful contributions to location theory have tended to "neutralize" the space in which their theories operated. That this generalized Euclidean space concept makes locational relationships easier to perceive and understand will not satisfy the city planner, however, for the critical quality of the space with which he deals is its variability, its possibilities and limitations; and the "limitless plain stretching endlessly in all directions from a point" is an abstraction of doubtful value to him. Thus, at the very beginning the most useful level of abstraction needs to be determined; a policy model requires a space "translatable" into real situations. At the same time, it must express spatial relationships with some ease; for among its basic inputs is the spatial arrangement of such items as employment centers and transportation systems, while the output is similarly a spatial arrangement of values and densities. Hence, the model's space must be quantifiable, as a physical commodity; capable of expressing locational characteristics in simple form; and adaptable to "real" situations with comparative ease.

The population characteristics offer a different kind of problem—what aspects of the urban population should we be talking about? "Persons" includes children, institutional populations, and others whose economic behavior is not relevant to the model (which is concerned with decision units), so it is necessary first to define the demographic

---

[25] Hereafter, the term "density" will be used to express "persons per unit area of space" as well as the inverse "area per person."

unit most useful to the model and second to specify what population compositions are acceptable.

The Bureau of the Census defines a "household" as "all the persons who occupy a house, an apartment or other group of rooms, or a room that constitutes a dwelling unit."[26] Since the household is the basis of demand in the market for shelter space, its behavior does affect the conclusions of the model and is the appropriate demographic unit for the model.

We know that different types of household display different kinds of behavior. High income households tend to behave like other high income households and quite differently from low income households. Racial minorities tend to cluster together, if only because alternatives are barred to them. Households with school age children are moved by quite different considerations than are childless households when making locational decisions. Taking all households, there is an extensive variety of locational behavior observable. Furthermore, in American cities demographic composition varies widely. All these things present difficulties in determining a useful composition of the household population. Accordingly, the model will be based initially on a set of households which is homogeneous in income, in the manner in which space is valued, and in the manner in which time is valued. This abstract set of households performs two functions that are essential to the model: (1) it produces a labor service which must be used at production sites, so that certain characteristics of its behavior in the labor market are pertinent to the model; (2) it purchases the services of urban land indirectly in seeking to optimize its location, so that certain consequences of its behavior in the market for shelter space are germane to the model. At a later point a more complex population of households will be manipulated in the model to illustrate effects of compositional variations.

The spatial and demographic features of the model must be brought together by a description of the processes which relate to them. Thus we will look at the city as an organization for the production and consumption of goods and services. The spatial arrangement of the city—its pattern of land using activities—is an expression of its organizational features and its technology, which involve bringing together at certain places in the urban space the labor services of the population, capital in the form of plant and equipment, and the appropriate physical inputs.

---

[26] 1950 Census of Population.

The model concentrates on how the labor services are organized in space, given the characteristics of the transportation system, the spatial arrangement of production, the nature of the labor force, and the institutions by which the labor force is articulated with the processes of production. The manner in which labor performs in production derives from social organization: the arrangement of work shifts, the journey-to-work, and the degree of labor orientation of industry are behavioral characteristics which are important in the model. The technology of the transportation system enters the model through its effects on the transportation costs of the journey-to-work.

Given the characteristics of the four general aspects of the city discussed in the preceding pages, the construction of the model will proceed along the following lines:

First: a concept of transportation demand based on certain characteristics of the labor force and of the journey-to-work will be developed.

Second: a systematic general description of the transportation function, based upon its technological characteristics and its response to demand, will be described.

Third: a general transportation cost function will be elaborated to integrate time-based costs, distance-based costs, and overhead costs as they affect the decisions of the demanding unit.

Fourth: a system of location rents which result from the transportation cost function will be described, and this, with a discussion of the implications of a supply of space, will round out the picture of the supply elements encompassed by the model.

Fifth: the manner by which the individual household "demands" space will be developed, so that we can describe the demand conditions of the model.

The final step of the model is the bringing together of the supply and demand elements so that a spatial distribution of location rents and household densities is generated.

In subsequent chapters, implications suggested by the model and the prospects for application will be explored.

II

Characterizing Technology In Urban Transportation Systems

Transportation and Urban Spatial Organization

As the preceding chapter suggests, the central role in the model which this study will develop is played by the urban transportation system. The term "system" may connote a higher degree of organization and integration than is intended and should be viewed as referring simply to the total array of opportunities for the movement of persons and goods between points in the urban region. The system contains a number of more particular and more intricately organized subsystems—for example, transit systems, street systems (including the carriers that operate over them), and pedestrian systems, so that when the term "transportation system" is used herein, it is in this most general context and is specifically limited to the movement of persons within the urban region.

That the urban transportation system should be assigned this crucial role in a model whose purpose is to tell us something about the distribution of persons and the value of land in the urban region is certainly consistent with what we know about the impact of transportation innovation on the spatial organization of the city. In an across-the-board view of the larger American cities during the first thirty years (1920-50) of the "automobile age" it appears that the cities which grew the most during this period tend toward substantially lower gross densities than those whose major growth took place before the impact of the auto began to be felt (Figure 5). The 1960 Census of Population indicates not only that the population growth of the great metropolitan areas since 1950 has been absorbed by their suburbs, but that many of the central

23

cities have been emptying out their populations into the surrounding hinterlands, a phenomenon facilitated by the great new urban expressways to serve the automobile and the rising real incomes extending the range of automobile ownership. We are literally surrounded by current evidence of these impacts; it is the specific nature of the impacts that is difficult to bring into focus.

The organizing role assigned to urban transportation has, of course, been central to the field of urban land economics for half a century. Richard Hurd's classic *Principles of City Land Values* was published in 1903; in it he succinctly formulated the interrelationship of urban land values and the urban transportation system, a proposition scarcely modified in current writing on the problem.[1] The pioneering works of Weber and von Thünen[2] in the economics of location similarly brought transportation to the fore as the distributor of economic activities over the landscape; their positions have recently been extended and amplified by such students of locational economics as Lösch, Isard, and Dunn.[3] And so it is that transportation has become a primary element in the theories about how economic activities get distributed on the one hand and land values are developed on the other.

Our initial task is to determine what features of urban transportation are essential to the problem of understanding how households get distributed within the metropolis, how these features relate to other crucial aspects of urban organization, and how the relationships can be made into quantitative statements to be manipulated. We begin with the proposition that the relevant characteristics of the urban transportation system can be summed up in the cost characteristics of the system. This

[1] Richard M. Hurd, *Principles of City Land Values* (1903), later edition (New York: The Record and Guide, 1924). See also Robert M. Haig and Roswell C. McCrae, *Regional Survey of New York and its Environs*, Vol. 1, "Major Economic Factors in Metropolitan Growth and Arrangement" (New York: Regional Plan of New York and Environs, 1927); Richard U. Ratcliff, *Urban Land Economics* (New York: McGraw-Hill Book Co., 1949); and E. M. Fisher and R. M. Fisher, *Urban Real Estate* (New York: Henry Holt & Co., 1954).

[2] Alfred Weber, *Ueber den Standort der Industrien*, Pt. 1 (Tubingen, 1909), translated and edited by C. J. Friedrich as *Alfred Weber's Theory of Location of Industries* (Chicago: University of Chicago Press, 1928); and J. H. von Thünen, *Der Isolierte Staat in Beziehung auf Landwirtschaft und Nationalökonomie* (Hamburg, 1826).

[3] August Lösch, *The Economics of Location* (New Haven: Yale University Press, 1954); Walter Isard, *Location and Space Economy* (Cambridge: Technology Press, 1956); and Edgar S. Dunn, Jr., *The Location of Agricultural Production* (Gainesville: University of Florida Press, 1954).

requires that the unit of transportation services be identified, that the kinds of cost relevant to the problem be delineated, and that the processes by which these costs develop be described. This chapter will concern itself with the cost characteristics of the time spent by persons using urban transportation systems.

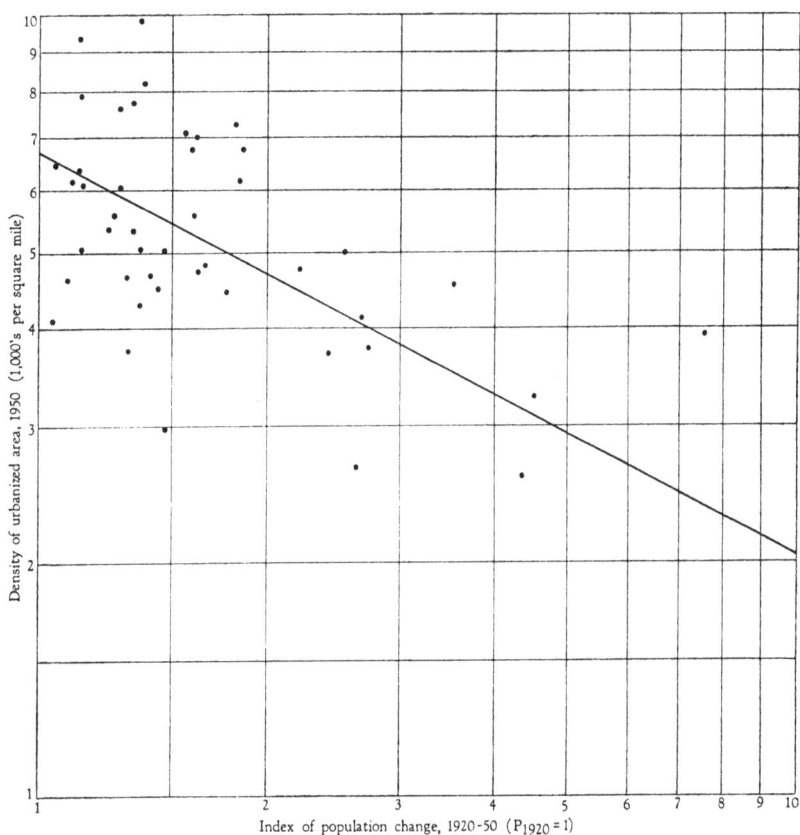

FIGURE 5. *Scattergram of 45 U.S. cities having (1) a population of 200,00 or more in 1950, and (2) a positive population growth during the period 1920-50 (relating index of growth, 1920 = 1, to population density of urbanized area in 1950).*

Source: *U.S. Census of Population, 1950*, Vol. I, Part 2: Growth, pp. 46-47, Table 23; Density, pp. 26-29, Table 17.

The Role of Accessibility

A parcel of "land" has two characteristics of interest to the economist: (1) a natural endowment, such as a class of soil, a mineral content, or water, for which society, given its technology and its wants, may have use; and (2) a quality of location with respect to the array of economic activities. The economic role of the endowment of urban land is generally confined to those topographical and geological features which affect its usefulness as site, or which condition the amount of "plant capital" necessary to bring it into a specified use. The quality of location, or "accessibility," is the dominant factor in determining the uses of the land and their intensity.

In a technical sense, accessibility is a relative quality accruing to a parcel of land by virtue of its relationship to a transportation system operating at some specified level of service. For purposes of an economic analysis this quality can be measured as the sum of the reciprocals of a specifically designated set of transportation costs associated with the parcel.[4] The "level of service" of a system should be distinguished from its "quality," which connotes a "level of amenity:" riding to work in a chauffeur-driven Cadillac as compared to car-pooling in an old Ford is using transportation of superior quality but not necessarily of a higher level of service, especially if both observe traffic regulations and endure the same traffic congestion en route. For our purposes the level of service of a transportation system is a reflection of the quantity of transportation services supplied by the system and the volume of demand that is asserted upon it: it is, in short, a measure of the system's efficiency, its "output" per unit cost. To build into a model the element of accessibility requires some analysis of the determinants of the level of service—the demand for and the supply of the services of transportation.[5] Later the

[4] Researchers using the potential model have followed a somewhat specialized definition of accessibility and have related it to the inverse of some parametric power of distance. Hansen, however, has replaced distance with time and thus more closely approaches an economic definition restricted to time costs. Cf. Hansen, *op. cit.*, p. *74*, especially fn. 2.

[5] "Transportation" here and henceforth refers only to the movement of persons, since the intraurban movement of goods involves only a small fraction of the total urban traffic—truck trips are of the order of 15% of the total vehicle movements in large cities. Cf. Chicago Area Transportation Study (hereafter referred to as CATS), Vol. I, p. 48, Table 7; and Detroit Metropolitan Area Traffic Study (hereafter referred to as DMATS), Part I, *Data Summary and Interpretation* (July 1955), p. 53. Outside of its contribution to congestion effects, truck traffic probably exerts little effect on the locational or transportation behavior of households.

"level of service" will be translated into a set of costs and hence, indirectly, into accessibility. From this point it is but a brief step to the principal problem—the intraurban distribution of households and land values.

The first step is to identify the "unit of output" of a transportation system. In fact, there is no single unit, and the unit chosen will vary with the purpose of analysis: the number of person-trips per period[6] is a useful measure of capacity; passenger- (or ton-) miles are an effective engineering measure for the output of a system; the simple aggregate of person-trips is an important measure of output when certain kinds of structural regularity are being sought in the data. Transportation yields a product which has three basic dimensions—volume, time, and space —and so it can be measured in different ways.

Take the case of the businessman in a large city who drives five miles to work on a Monday morning. At the most general level he has made one of several million person-trips made on that morning. Put in a time dimension, he ended his trip at approximately the same time that several hundred thousand other residents ended theirs, and so he was a part of the morning "peak flow." Put in a space dimension, he followed a route which was one of a limited number of reasonable alternatives that existed for him because of the manner in which the street system is laid out. Joining both time and space dimensions, we can see how his delay in a traffic jam resulted from the fact that a large number of other drivers were seeking to occupy the same space he was occupying at the same time that he was occupying it, and how accordingly his average velocity fell to only fifteen miles per hour compared to his normal twenty. It is apparent that the nature of the analysis will determine which of the dimensions of the unit are relevant and which can be held constant.

Concepts of Transportation Demand

The "demand for transportation services" is the number of units of output which will be consumed at various levels of "cost." Because the unit of output has several dimensions, the demand for these services has

---

[6] "A person-trip is defined as a one-way journey by a person travelling as a driver or passenger in an automobile, or as a passenger in a taxi, truck, or mass transportation vehicle, taking the person outside the block of trip origin . . . person-trips are linked trips. In a linked trip, a person using two or more modes of transportation to proceed from origin to destination is considered as making only one trip." CATS, Vol. I, p. 29, fn. 2.

several meanings also. Each concept of demand involves different conditions and responds to different variables: the conditions determining how many trips people will want to make are quite different from those determining how many people will want to shop downtown on Thursday morning. Five different concepts of transportation demand in urban systems can be identified and labeled: the "demand for movement" is concerned only with the aggregate number of trips that people in a given urban area will want to make; "traffic demand" is concerned with the number of trips in terms of their spatial characteristics; "load" deals with the distribution of trips in time; "flow demand" relates to the number of units that want to pass a specified point at a specified time; and where that point is the common destination for a number of trip movements, we will speak of "deadline demand."

"Movement demand" can best be understood in a purposive framework. People make trips between locations to engage in transactions of one sort or another, and the basic purpose of a transportation system is to facilitate the assembly of persons engaged in those transactions.[7] They may be simple money transactions involving the purchase of goods and services, or they may involve the exchange of labor for wages, as in the journey-to-work, or they may be transactions of a more personal sort, such as participating in a social or recreational experience, but virtually all movement is carried out for specific social and economic purposes.

The demand for movement is anchored in the communication requirements of urban society and its activities. Technological change alters these requirements, as do changes in social institutions, but an even greater source of change lies in the ability of society to substitute communication for transportation, to replace the movement of goods and persons by the transmission of information. To get some idea of the magnitude of movement demand in urban areas one can examine the aggregated data from any recent urban transportation study. The city of Detroit was found to require 5,263,975 internal person-trips per day, or 1.75 per capita per day, in 1954.[8] In 1957 the per capita daily trip figure for 1,417,000 persons in St. Louis was 1.76.[9] In the Washington

---

[7] A very small segment of the total volume of movement can be construed as having an end unto itself: that performed by the "joy-rider," the "Sunday driver," and the like.

[8] DMATS, p. 34, Table 9, and p. 57, Table 16.

[9] St. Louis Metropolitan Area Highway Planning Study (St. Louis, 1959), Vol. I, *Highway and Travel Facts,* Summary, 4th page; and Vol. II, *Tables and Reference Data,* p. 19, Table B-2.

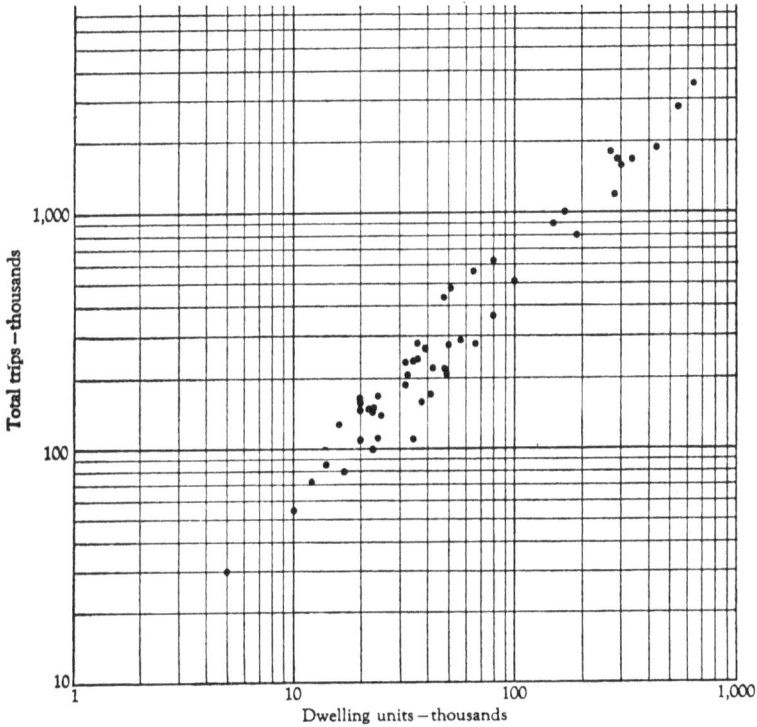

FIGURE 6. *Number of trips related to number of dwelling units.*
Source: Frank B. Curran and Joseph T. Stegmeier, "Travel Patterns in 50 Cities," *Public Roads—A Journal of Highway Research* (Washington: U.S. Department of Commerce, Bureau of Public Roads), December 1958, Vol. 30, No. 5.

metropolitan area the total volume of movement grew from a total of 2,000,000 daily person-trips, 1.79 per capita, in 1948 to 3,140,000, or 2.0 per capita, in 1955.[10] In Chicago in 1958 the total was 9,931,000 internal person-trips per day, or 1.92 per capita.[11] One should, of course, avoid any confusion between "demand" and an observed volume,

[10] Mass Transportation Survey, National Capital Region, *Traffic Engineering Study* (Washington, 1958), p. 25.
[11] CATS, Vol. I, p. 108, Table 19, and p. 117, Table 30. "If the total number of *resident* trips with destinations in the Study Area is divided by the number of persons, the figure of 2.0 trips per person per day is obtained. This is equivalent to one round trip per person on the average week day. *So, if a million persons are added to the population of the Study Area, at least two million additional trips will be made each week day."* (Italics mine.) CATS, Vol. I, pp. 31-32.

but the concept of demand is implicit in the per capita figures, for these suggest that in large cities purposeful trips are required at a rate of from 1.75 to 2.00. Curran and Stegmeier's study of the traffic characteristics of 50 cities supports this relationship, but it goes further, in that it suggests that the trips per household tend to decline with the size of the city.[12] Thus, as cities grow in population, we can expect the total volume of movement to expand, although there is good reason to expect that the per capita requirements will decline under the impact of technological change and of greater efficiency of movement of larger cities (Figure 6).

"Traffic demand" is used here in a special sense: it is the demand for movement viewed in the light of the geography of origins and destinations of its trips. The traffic demand of work trips, for example, comprehends the "desire lines" of travel for all workers between their homes and their employment sites and hence is concerned with the twin distributions in space of employments and dwellings. If the work trips are screened out and their desire lines plotted, two major spatial classes of movement emerge: radial movements from outlying residential areas to and from the employment centers in the urban core, and circumferential movements between domiciles and employment sites which are not located at the core. Historically, the core movement has been by far the greatest, and this suggests that traffic demand generated by the work trip can be viewed in a radial-concentric spatial framework without doing violence to the gross dimensions of the spatial characteristics of the work trip. If employment continues to decentralize and shift outward from the core, the circumferential movements will tend to become relatively more important, and the simple, concentric spatial framework will require modification.

If the demand for movement is distributed over time, the "load" characteristics of urban transportation are revealed. Plotting trip data by work and nonwork classes and by hourly volumes over the 24-hour day reveals the characteristic bimodal load pattern of urban transportation; but, further, the dominance of the work trip in setting the levels of demand throughout the system emerges in dramatic fashion (Figure 7). Here we can apply an intellectual tool originating in the engineering of uitlity systems to characterize the concentration of demand in time. "Load factor" is the ratio of the average to the peak, or maximum,

[12] Frank B. Curran and Joseph T. Stegmeier, "Traffic Patterns in 50 Cities," *Public Roads—A Journal of Highway Research* (Washington: U.S. Department of Commerce, Bureau of Public Roads), December 1958, Vol. 30, No. 5.

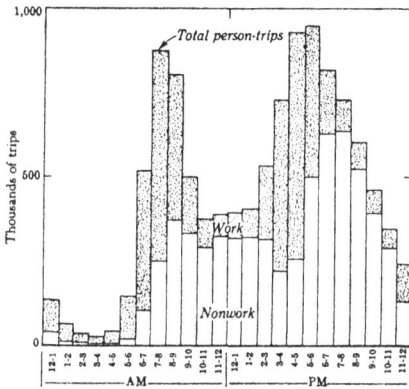

FIGURE 7. *Hourly distribution of internal person-trips by work and nonwork purposes, Chicago, 1959. (All trips to home assigned to work or nonwork categories on assumption of 9-hour lag between home-to-work and work-to-home trips.)*
Source: See Figure 8, below.

volume; a low load factor (approaching zero) reflects a high degree of concentration of demand, while a high load factor (approaching one) describes hourly volumes among which there is little variation. Taking the volume of trips as a whole, urban transportation systems may exhibit fairly high load factors—the Chicago data suggest a load factor of almost .500. Figure 8 demonstrates a characteristically low load factor of .294 for work trips and a somewhat higher one of .413 for nonwork trips. Because the work trip peaks are slightly offset with respect to the

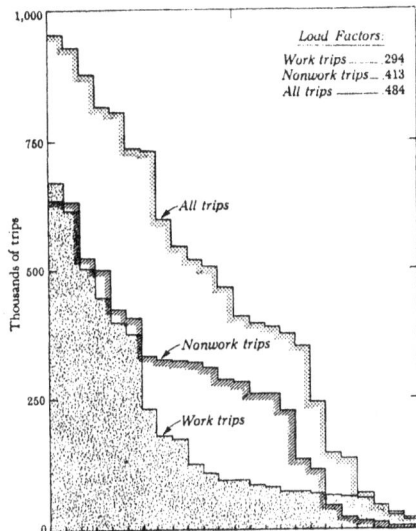

FIGURE 8. *Load diagram, all internal person-work trips, by hours for 24-hour period, Chicago, 1959.*
Source: Chicago Area Transportation Study, Vol. I, *Survey Findings* (Chicago, 1959), p. 35, Fig. 15.

nonwork trip peaks the two lower load factors result in a higher aggregate load factor for the system.

Finally, the distribution of movement in both time and space provides a concept of demand which is most useful for this study—"flow demand." Transportation flows can be analyzed in terms of the "orderliness" that is imposed on certain classes of movement by the institutional requirements of society. The patterns of movement for certain purposes display strong regularities because trips for those purposes are rigorously ordered in time and space;[13] the patterns of other movements are almost random because their purposes can be realized in a number of ways, at a variety of places, or at any convenient time.

Recent traffic studies have developed "trip purpose" data which highlight some of the regularities in urban transportation that emerge from the institutional aspects of society.[14] A general correspondence between the degree of orderliness and the relative volumes attributed to trip purposes is noteworthy. Thus, Curran and Stegmeier's data for fifty cities indicate that 55.8% of all person-trips are between pairs of purposes, at least one of which is "work and business."[15] This was supported in Detroit, with a figure of 51%,[16] and in Chicago, with a slightly higher one of 56%.[17] In both Detroit and Chicago the trips between home and work constituted almost exactly one third of the total trips tabulated. Although not directly comparable to the Chicago and Detroit data because of the lumping together of work and business purposes, the visual presentation by Curran and Stegmeier of their data in a "purpose-to-purpose matrix" affords a valuable perspective of the relative importance of movement elements (Figure 9 and Table 1). The principal conclusion of these data is the overwhelming importance of the journey-to-work in the aggregate demand for movement.

The basic "generator" of person movement is the household. According to a study of origin-destination data from 38 cities by the Bureau of

[13] When an element of movement is repetitive, taking place regularly and frequently, and when both the origin and destination tend to remain spatially fixed in the short run, we will say that it is an "ordered" element.
[14] Especially revealing are the tabulations of trips by purpose of origin and of destination. Cf. DMATS, p. 124, Table 37; CATS, p. 37, Table 4; St. Louis Metropolitan Area Highway Planning Study, Vol. II, p. 19, Table B-2; and Curran and Stegmeier, op. cit., p. 120, Table 13.
[15] Curran and Stegmeier, op. cit., p. 120, Table 13.
[16] DMATS, p. 124, Table 37.
[17] CATS, p. 37, Table 4.

TABLE 1.—*Number and Percentage of Trips in 50 Cities from Each Purpose to Each Purpose, All Modes of Travel*

			Trips from—			
Trips to—	Work and business	Social and recreation	Shop	Miscel-laneous	Home	Total
Work and business						
Number	1,374,091	61,684	69,598	626,348	5,627,591	7,759,312
Per cent	4.9	.2	.3	2.3	20.2	27.9
Social and recreation						
Number	146,943	476,713	108,890	197,421	2,406,844	3,336,811
Per cent	0.5	1.7	.4	.7	8.7	12.0
Shop						
Number	178,461	96,230	234,763	145,471	1,435,139	2,090,064
Per cent	0.7	.3	.8	.5	5.2	7.5
Miscellaneous						
Number	633,717	171,409	97,351	453,892	1,925,011	3,281,380
Per cent	2.3	.6	.4	1.6	6.9	11.8
Home						
Number	5,414,774	2,533,006	1,570,604	1,791,704	......	11,310,088
Per cent	19.5	9.2	5.6	6.5	......	40.8
TOTAL						
Number	7,747,986	3,339,042	2,081,206	3,214,836	11,394,585	27,777,655
Per cent	27.9	12.0	7.5	11.6	41.0	100.0

Source. Frank B. Curran and Joseph T. Stegmeier, "Traffic Patterns in 50 Cities," *Public Roads—A Journal of Highway Research* (Washington: U.S. Department of Commerce, Bureau of Public Roads), December 1958, Vol. 30, No. 5.

Public Roads, the average household originated approximately one work trip each working day, one business trip every seven working days, one social recreation trip every other working day, one shopping trip every fourth working day, one school trip every tenth working day, and a miscellaneous trip every fourth working day.[18] Even at this level, the dominance of the journey-to-work is obvious. Social and recreation trips form the next most important class. The destination of such a trip may be any one of a multitude—to the home of a friend, to the site of some

[18] Robert E. Schmidt, and M. Earl Campbell, *Highway Traffic Estimation* (Saugatuck, Conn.: Eno Foundation, 1956), p. 20 and Table II-4 consolidating trip purpose data for 38 cities.

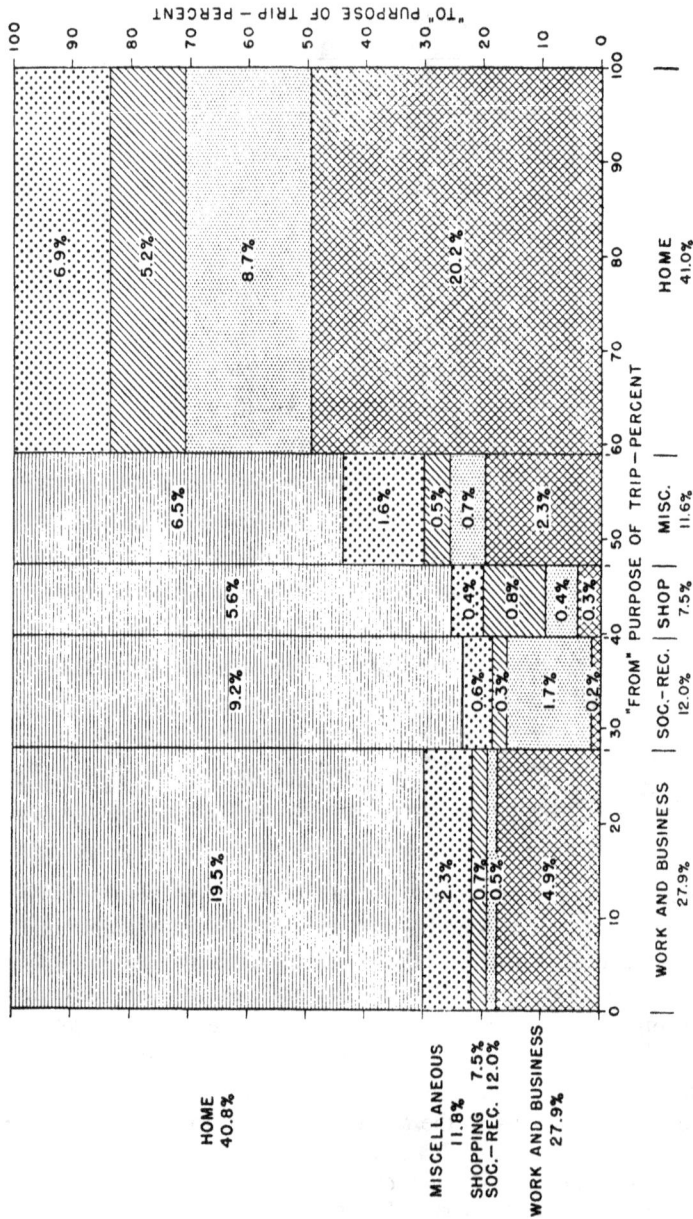

FIGURE 9. *Percentage distribution of trips from each purpose to each purpose (journey-to-work sectors emphasized).* Source: Frank B. Curran and Joseph T. Stegmeier, "Traffic Patterns in 50 Cities," *Public Roads—A Journal of Highway Research* (Washington: U.S. Department of Commerce, Bureau of Public Roads), December 1958, Vol. 30, No. 5.

commercial recreation, to a country club or park. It may be made by any member of the household and at almost any time, except that such trips tend to be somewhat concentrated in the evenings and on the weekends. The shopping trip is more orderly. Convenience items, such as food and gasoline, are purchased regularly and at locations which, if not predetermined, are at least delimited by spatial relationship to the household. At the other extreme are specialized goods and services which may be purchased rarely, and by few households, and for which consumption patterns are, hence, completely irregular. As a rule the more frequently it is performed by the household, the more the transaction will tend to be regularized in the time and space distribution of movement; at the same time, the dominant types of shopping movement are those most likely to have some regularity. School movements, like the journey-to-work, are ordered by the schedule of school activities, but since they constitute less than 5% of the total dwelling-originating movement, they are inconsequential. A randomness can be ascribed to the miscellaneous class. In general, it is possible to estimate from these data that something over half of the total dwelling-originating movement is highly ordered, that perhaps one third is largely disordered in time and space, and that the remainder is partially ordered.

The journey-to-work—the technological link between the labor force and the production process—is the most significant class of person movement in any urban region, in terms of order and relative volume. In addition, the journey-to-work probably has the lowest "price elasticity" of demand of all of the classes: it would take a very large change in the cost of the work trip to have a perceptible effect on the number of person-trips in the short run.[19] In the aggregate, the journey-to-work has a higher degree of order than the other classes of movement because of the manner in which work is institutionalized in the community. The other classes of movement relate to activities whose location tends to be dependent on the distribution of households or to situations in which the household has a considerable number of alternatives as to where and when its transactions may be carried out. Finally, the rigorous time constraints of the journey-to-work are responsible for the massive "peaking" in urban areas—and, hence, the low load factors—of the demand for transportation services.

These considerations suggest that little violence will be done to our

---

[19] Although substantial shifts might occur among modes of transportation, or even within modes, as in the case of car-pooling practices.

conclusions if we concentrate on the home-work relationship and construe all other transportation classes to play an indeterminate role in household location and transportation behavior. For the purposes of this analysis, then, the demand for movement is simply the product of the employed labor force and the frequency of the work periods—the total number of work trips during a period of time necessary to support the processes of production. Characterizing traffic demand requires that the distribution of employment and households in the urban space be identified. Defining the loading of the system involves defining the time characteristics of work trips—their periodic nature, their duration, and their simultaneity. When we concentrate on the journey-to-work and bring together the load and traffic demand factors into a common frame, a highly specialized type of flow demand emerges—the "deadline demand" imposed upon the transportation system by the institutional manner in which economic activities are carried on.

Capacity—The Supply of Transportation Services

Although there are several useful ways to define the demand for transportation services, this ambiguity does not exist for "supply;" "the supply of movement" has little meaning except as a general description of the physical plant available for movement, unless it refers specifically to the *opportunities* for movement between designated points in space at specified times. Fundamentally, this is what is meant by "capacity."

When the scope of analysis is narrowed by the exclusion of all components of movement but the journey-to-work, there remains a transportation system having several special characteristics:

(1) Spatially concentrated employment sites and the dispersed domiciles of the labor force are linked by transportation routes over which carriers of several sorts operate.

(2) The movement of persons over the transportation facilities is highly concentrated in time, the result being a diurnal cycle typified by peaks of demand immediately prior to and following the work periods.

This combination of temporal and spatial concentration inherent in the journey-to-work is the dominant feature of urban transportation. Given the capacity characteristics of such a system, the peaks of demand will necessarily result in the system's operating under conditions of saturation—where no additional carriers could be added without their delaying those following—for at least some period of time. Basically this is the "uniform speed" condition which Martin Beckmann[20] described as resulting from the increase in flow and decline of passing opportunities in a "free speed" situation. At some point as flow increases, the individual carriers lose freedom to choose velocity and become prisoners of a situation in which the range of decision is reduced to adapting to changes in the flow.

To understand better the relevant characteristics of a saturated transportation system, it is useful to carry the problem to a more abstract level. First, we will focus on only one of the many routes linking employment sites and dwelling places, since we can generalize from it to the others. Second, we will assume a homogeneity of carriers in all relevant features to sidestep problems of "mix." One could assume a system based on rail transit, buses, or pedestrians by appropriately describing "capacity" in terms of that system, but for this analysis focus on the private passenger automobile is suggested by several considerations: its relative dominance as the carrier in the work trip, the importance of congestion effects resulting from the peaking of auto work trips, and the availability of a body of analysis of auto traffic characteristics. Third, we will let all trips be constrained by a common arrival deadline and by a common time of discharge. This abstraction will be referred to as a "movement system."

Capacity in a uniform flow system is determined by the velocity at which the carriers move and the distance between them.[21] It follows that if they could move in a system so that the interval between them were independent of their velocity, the volume handled by the system could be increased by an increase in the system's velocity, and capacity would be limited only by the technological limits to the velocity of the carriers. In reality, however, where each unit is controlled independently,

---

[20] Martin Beckmann, C. B. McGuire, and Christopher Winsten, *Studies in the Economics of Transportation* (New Haven: Yale University Press, 1956), pp. 32-33.

[21] The *Traffic Engineering Handbook* expresses this as

$$c = \frac{5280V}{S}$$ where c = velocity-specific capacity in vehicles per hour,

$\quad\quad\quad\quad V$ = velocity in miles per hour, and

$\quad\quad\quad\quad S$ = the spacing (interval) between successive vehicles in feet.

Traffic Engineering Handbook, 2nd ed. (New York: Institute of Traffic Engineers, 1950), p. 331.

as in the case of automobile traffic, the interval between carriers is the "safety factor" necessary to avoid collision and interruption of flow and is directly related to the time required for a carrier to accommodate its behavior to that of the carrier ahead. If this time is constant, the distance between units is directly related to their velocity, and the amount of space in the moving stream of traffic required by a unit will vary with the velocity of the flow. Consequently, the number of units that can be accommodated per unit-time, that is, the capacity of the system, depends directly on the flow velocity.

Here two senses of capacity should be distinguished:[22] (1) for any given speed a flow capacity is reached when no space in the flow is wasted, a condition designated "velocity-specific capacity;" (2) "system capacity" is the maximum velocity-specific capacity for all speeds of flow, and the velocity at which it is reached is labeled the "capacity velocity." It is capacity in this latter sense that is to be understood hereafter, unless otherwise qualified.

The binding limit to the capacity of any system in which control is decentralized is imposed by the interval required to maintain uninterrupted flow, and the size of this interval is directly related to certain behavioral characteristics of the system:[23] the mechanical characteristics of the carriers, the responsiveness of control (or reaction time), and the level of risk at which the system operates. The risk factor is related to the "orderliness" of the system, basically the degree to which control in the system is centralized (the ultimate in an orderly system would be a moving conveyor belt). The kind of system described above would be completely orderly if the units could anticipate each other's behavior

---

[22] The complexity of the concept of capacity has given rise to a multiple definition in highway traffic planning:

"*Basic capacity:* the maximum number of passenger cars that can pass a given point on a lane or roadway during one hour *under the most ideal roadway and traffic conditions* that can possibly be attained. . . .

"*Possible capacity:* the maximum number of vehicles that can pass a given point on a lane or roadway in one hour *under prevailing roadway and traffic conditions.* . . .

"*Practical capacity:* the maximum number of vehicles that can pass a given point on a roadway or designated lane during one hour without *traffic density being so great as to cause unreasonable delay, hazard, or restrictions to maneuver under prevailing roadway and traffic conditions.* . . ."
Bureau of Public Roads and Highway Research Board, *Highway Capacity Manual* (Washington-GPO, 1950), pp. 6-8.
See also *Traffic Engineering Handbook,* p. 334.
[23] But *not* by the actual behavior of the units.

with sufficient "lead time" to avoid collision, and the interval would then be limited only by the size of the carriers. In a disorderly system the behavior of the lead carrier is completely unpredictable; the behavior of the second carrier as it accommodates itself to the first is unpredictable to the third, and so on through the system. Hence, capacity is related to the level of disorder, and if there is any degree of disorder in the system, reaction time and the mechanics of the system will set the "safe" interval.

"Reaction time"[24] refers to the time interval between the perception of a change in the velocity of the carrier ahead and the assertion of controls to adjust to it. Again, if this reaction could be made instantaneous and if the mechanical characteristics of all carriers were uniform, the interval could be reduced to zero. The human response is largely a neurological one whose duration, invariant with respect to the analysis most traffic authorities take to be about one second. The second factor influencing reaction time, the mechanical ability of the carrier to change velocity, is related to certain technological variables: the mass of the carrier, its braking power, and its velocity.[25]

The general form of the mathematical expression for the length of the interval between carriers[26] can be stated as:

---

[24] In a vehicular traffic system, the traffic engineer characterizes this as "driver perception and reaction time."

[25] This relationship has been generalized for automobiles in the *Traffic Engineering Handbook,* p. 72, as follows:

$$B = \frac{.0105 \; V^{exp \; 7/8}}{f}$$

where B = vehicle braking distance in feet

V = velocity in miles per hour, and

f = coefficient of friction (at a velocity of 20 mph.)

which can be restated

$$B = \frac{.0105 \; N V^{exp \; 7/8}}{F}$$

where F = the minimum force necessary for deceleration, and

N = the mass of the carrier.

[26] The *Highway Capacity Manual* (p. 3, Table I) presents some 23 expressions for "safe following distance" in terms of velocity. All of these expressions have the general form:

(1) $I = av^\beta + \gamma v + \zeta$

where a and β are parameters for the mechanical characteristics of the
    carriers,
    γ is the reaction time,
    ζ = the length of the carrier,
    v = velocity, and
    I = the interval between the corresponding points of
        successive vehicles in the traffic flow.

From this the general form for the capacity of a decentralized movement

$$I = aV^\beta + \gamma V + L$$

It we let $a = \dfrac{aMr}{F}$,

                                  $\beta = b$, and

                                  $\gamma = rt_c$,   .

This general form is converted to our own expression, which is the denominator in
Equation 2 in text. Thus, "safe following distance" corresponds in mathematical
expression with our concept of interval, especially if we let r = 1, as seems
indicated. (See Equation 2 below.)

If we disregard the five expressions which accept $a = O$ (which appear to
assume no braking distance), and the three expressions which take $\gamma = O$ (thus
assuming no driver reaction time), we are left with fifteen expressions which are
"full." These fifteen equations describe a set of curves which are convex upward
and which are heavily skewed in the direction of the vertical (or capacity) axis.
(See *ibid.*, p. 2, Figure 1.) Their maximum speeds fall between 8.3 miles per
hour and 28.3 miles per hour, and they range in volume between 1,050 and 2,520
vehicles per hour for a single traffic lane. The values for t_c range from .50 to 1.00
second, with nine accepting the high value; thirteen of the fifteen accepted $\beta = 2$,
while two assume $\beta = 2.3$ (this is the decimal equivalent of 7/3)—see footnote
25; the values for a when divided by the conversion factor $\dfrac{(22)^2}{15}$ to compensate
for expression in terms of ft/sec. range from .0116 to .157, with twelve using
values less than .05. The important point is that *as long as a, β, and γ have
positive values greater than 1, there will always be a "maximum" capacity associated with a system, and that capacity will be defined by the velocity. It is this
"maximum" capacity that is implied in our use of the term "system capacity."*

It should be acknowledged that where feedback is instantaneous, and where
the mechanical characteristics of all carriers are uniform, theoretically the most
persuasive expression is that which takes the coefficient $a = O$, such that
                        $I = \gamma V + L$
A little reflection will justify this: if the feedback time is instantaneous two
vehicles could be moving in the system "bumper to bumper" ($I = L$) and any
change in the behavior of the first would be simultaneously compensated for by
the second; here both a and $\gamma = O$.

Where the braking distance is uniform for all units, it cancels out as a spacing
element, leaving only the independent variables of velocity and feedback time.

system can be derived,[27] based on the general interval expression modified for the risk factor:

$$(2) \qquad C = \frac{V}{\rho(a'V\beta + \gamma'V) + \zeta}$$

A capacity expression based on such a spacing expression yields a C that increases with V to some asymptotic value. This value is, in fact, γ^{-1} at a capacity velocity of infinity.

For purposes of simplicity, this books assumes that γ has a positive value greater than O, because this appears to conform more closely to empirical observations. See also discussion in Beckmann, et al., *op. cit.*, p. 19.

[27] The relationship of capacity to velocity in decentralized systems can be developed in somewhat greater detail.

The velocity-specific capacity c of the system is a direct function of the velocity V of the system and an inverse function of the interval I among units:

$$(1) \qquad c = \frac{V}{I}$$

Interval contains three components: the length of the unit L, the interval component resulting from reaction time i_f, and the interval component associated with the mechanical characteristics of the system i_m, the latter two being modified by a disorder function, henceforth expressed as "risk coefficient" r as follows:

$$(2) \qquad I = L + r(i_f + i_m)$$

Here the risk coefficient r will vary from zero (where the system is completely orderly) to 1 (where the system is completely disorderly).

We can now give mathematical expression to the relationship between capacity and the factors associated with it.

$$(3) \qquad i_f = Vt_c$$

where t_c = the time lapse associated with the feedback organization.

$$(4) \qquad i_m = \frac{aMV^b}{F}$$

where a and b are technological constants,

$\quad M$ = mass of carrier,

$\quad F$ = minimum braking power to decelerate a carrier of mass M moving at velocity V to $V = 0$ in distance i_m.

$$(5) \quad \text{Then} \qquad c = \frac{V}{rV\left[t_c + \dfrac{aMV^{(b-1)}}{F} \right] + L}$$

(Substituting 3 and 4 in 2, and 2 in 1)

$$(6) \qquad \text{and } C = c(max) \qquad C = \text{system capacity}$$
$$= c, \text{ where } dc/dv = 0$$
$$\text{and the second derivative} < 0$$

where C = maximum capacity of a single lane operating under saturated conditions, [28]

ρ = risk coefficient $(0 \leqslant \rho \leqslant 1)$,

$a' = a/\rho$, and $\gamma' = \gamma/\rho$,

V = the capacity velocity [where $C = c(\max)$, $\dfrac{dc}{dv} = 0$,]

and $\dfrac{d'c}{dv} < 0$.]

and β = 2 or 2.333 for automobiles.

In general, such an expression is adaptable to most forms of carriers, but the capacity expression is in terms of the number of carriers which can pass a point in the period of time set by the velocity statement. It relates to the number of persons moving in the system only if there is a one-to-one correspondence between carriers and persons. If the number of persons per carrier is greater than one, the expression on the right would have to be multiplied by the average vehicle occupancy. Likewise, if a route consists of more than one lane in the direction of movement, the expression on the right would have to be multiplied by the number of lanes. And so, the carrier capacity of a movement system will vary with the velocity of the flow, as determined by the level of system disorder, by the nature of the control system, and by the mechanics of the carriers in the system. The trip capacity of the system will involve additional variables—the number of route lanes and average vehicle occupancy.

A final important feature of the carrier capacity of a movement system is that it is measured at the point of least capacity in the system. Such a point, or "choke," sets the level of flow for the entire route. The capacity choke explains the development of certain kinds of congestion within the transportation system, as will be discussed at a later point.

---

[28] For auto traffic, "safe (velocity-specific) capacity" is that relationship between velocity and interval that exists when the spacing of units is governed by "driver stopping distance." Since auto traffic has the characteristics of a highly disordered system, it has a risk coefficient approaching 1. Thus, the traffic engineer's definition of driver stopping distance is analogous to our concept of interval with a risk coefficient of 1. Driver stopping distance is defined in the *Traffic Engineering Handbook* as having three components (p. 71):

Driver perception-reaction time (which is analogous to our feedback component),

Brake-lag distance, and

Vehicle braking distance (the latter two being subsumed in the mechanical component of our formulation).

Time-Costs Defined

To contend that time is a commodity valuable to human beings should startle no one. The classical theory of the labor market is based upon the proposition that the worker "rents" his skills to the employer at some wage not less than the subjective value to him of the time surrendered. Transportation engineers, indeed, have included in their calculations as a cost factor an allowance for the value of time spent in transportation and have considered among the benefits of projects the value of time saved.[29] How time and money costs ought to interact is a subject for later discussion; how time-costs arise in transportation systems needs to be fully understood before the valuation problem is relevant.

In the analogy between time and money costs, the similarity of the simple cost equation and the equation in physics for uniform motion is immediately apparent:

$$c = pq, \text{ and } .$$
$$t = v^{-1}s$$

If we construe distance s as the amount of space "consumed" and as analogous to quantity q in the cost formula, then t represents not just time but a "time-cost" analogous to money costs c, and the price at which space is "consumed" or overcome is represented by the inverse of the velocity v. In this comparison, both the price p in the cost expression and the velocity v in the motion expression are to be viewed as externally set parameters. Finally, both are "constant cost" expressions: no matter what value q or s takes, the average cost does not change for the individual.

The motion equation viewed as a time-cost expression has an obvious application to transportation systems. Given the effective distance between an origin and a destination, under conditions of uniform velocity t will represent the time-cost of a single, one-way trip. Thus, where there is nothing to impede movement—that is, where an individual is free to choose and hold to his optimum velocity in consummating the trip—the motion equation affords a measure of the time-costs that will be borne by the trip-maker.

In fact, however, the conditions allowing travel at a uniform velocity will rarely obtain in an urban transportation system. The freedom to

[29] See American Association of State Highway Officials, *Road User Benefit Analyses for Highway Improvements* (Washington, 1955); also Schmidt and Campbell, *op. cit.*, p. 141.

choose and maintain a desired rate of movement is impaired by the fact
that any carrier shares facilities for movement with other carriers whose
behavior restrains his own. The degree to which one must depart from
a desired velocity depends on the density of the traffic through which he
must move; impediments range from a small deficit in passing opportunities to the saturated flows typical of urban congestion, where "freedom of speed" is completely abrogated. Pursuing the analogy above,
the reduction in velocity v is in reality an increase in the "time-price"
v^{-1} of the distance traversed. Unless the reduction in v is uniform over
the route—a special case at best—the "constant cost" characteristics
are lost, and marginal costs will tend to vary independently of s.

These points can be illustrated with a motion diagram (Figure 10).
Here OD represents the route between the origin and destination of a
trip, the horizontal axis representing the distance coordinate generally.

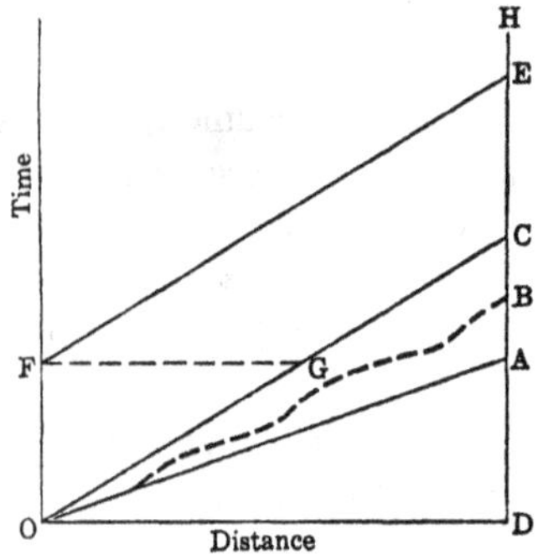

FIGURE 10. *Motion diagram.*

The vertical axis measures the time duration of a trip between O and
DH. Hence the slope of any line measures the marginal time-cost v^{-1}
of the movement at the point indicated. The "constant cost" case is
represented by a line of constant slope from O to DH. If OA represents the velocity desired for an individual to travel from O to DH,[30]
the total cost of the trip is represented by AD, where his movement is
not impeded. Where motion is impeded infrequently and in a random
manner, the irregular line OB might represent the movement, with BD
measuring the time-costs. AB, then, represents the net time-costs resulting from the interruptions. Finally, OC represents a uniform decrease

[30] The slope will always be positive, since negative velocity and negative time-
costs are meaningless.

in velocity from *OA* or a higher level of constant costs. In short, the decline in average velocity is identical with an increase in time-costs and is represented by an upward shift in the slope of the motion vector. Where freedom of behavior is completely lost in a saturated system, the individual is shifted from his desired vector *OA* to a new vector *OC*, and uncompensated time-costs in the amount of *AC* are imposed on him. It is not only the upward shift in the slope that is costly, however. Where saturation results from the influence of a deadline demand situation, as in the case of the journey-to-work, the individual's departure time must be chosen to assure meeting the deadline. If *E* is the arrival deadline and *FG* the length of the flow demand generated at *D*, then *FE* is the locus of the last unit which can meet the deadline, and *OC* that of the first unit to arrive; then *FO* = *EC* is the *maximum ingression loss* for any individual in the flow.

At this point our considerations of the nature of supply and demand conditions in urban transportation systems converge. In its simplest form "ingression"[31] is a condition in movement systems which results when the *instantaneous* demand on the system exceeds its *instantaneous* capacity, that is, when the number of units required to arrive at a given point, such as an employment center, is greater than the ability of the system to admit them simultaneously. Its primary characteristic is the queue, and its essential consequence is a time loss to each unit which depends on (1) the velocity with which the queue moves (or the capacity of the system), and (2) the position of the unit in the queue.

This construct is directly relevant to circumstances surrounding the journey-to-work. The fact that the assembly and dispersal of employees is governed by deadline conditions causes massive instantaneous demands upon the transportation system. At the same time instantaneous capacity is limited—for automobiles, to the number of "lanes" that are available. Each lane can admit one carrier each c^{-1} minutes, where c stands for the capacity of the lane. Thus the fundamental relationship between supply of and demand for the services of an urban transportation system can be summed up in the concept of ingression. Where n stands for the number of units making a deadline demand upon a system, c its capacity, Y the total time loss to all units in the queue, and

[31] For a more complete discussion of the ingression phenomenon, see Lowdon Wingo, Jr., "Measurement of Congestion in Transportation Systems," *Planning and Development in Urban Transportation—1959,* Highway Research Board, Bulletin 221 (Washington, 1959) pp. 1-28.

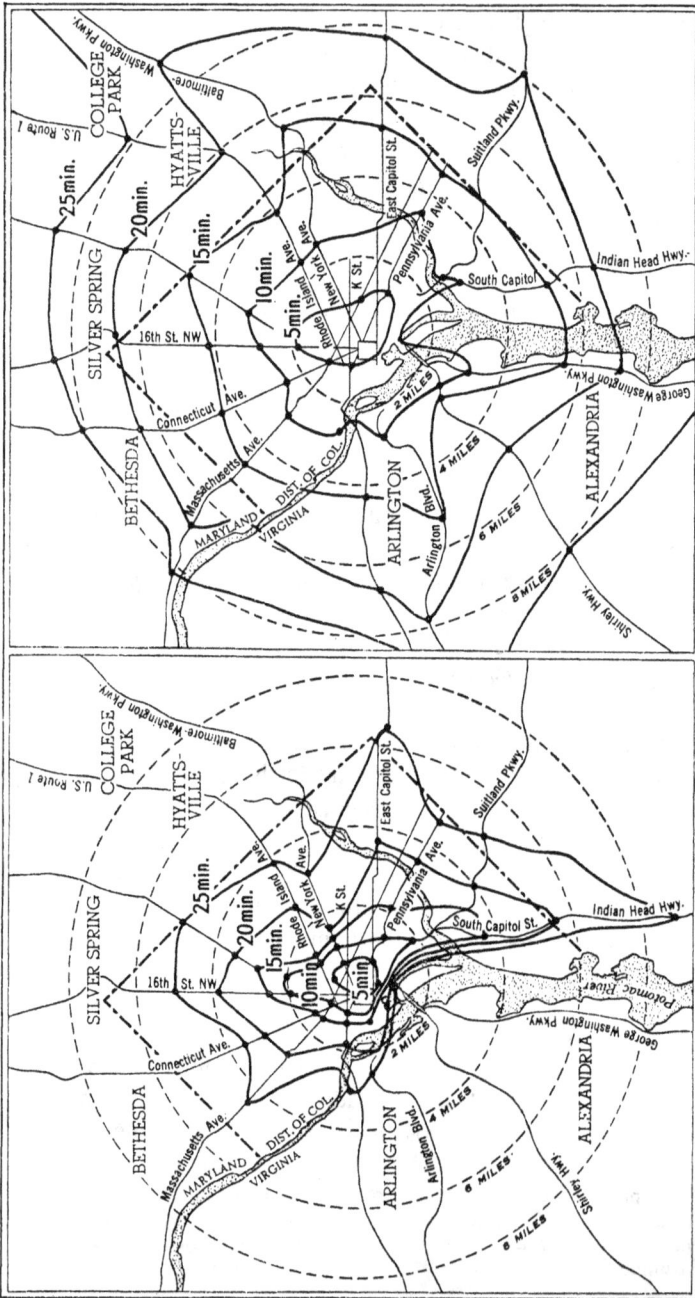

FIGURE 11. *The heavy lines on the map to the left reflect the snail's pace at which motorists progress in getting out of downtown Washington at the peak of the evening rush hour, or about 5:15. Right is a map of the same area with heavy rings to indicate the faster pace an hour or so later. Data for the maps were provided in a survey by the District Highway Department in which drivers reported their positions every five minutes during the test runs in the rush hour and post-rush hour periods.*

Source: Washington Post, December 3, 1958.

y_n the loss to the nth unit in the queue,

(3) $$y_n = \frac{n-1}{c}, \text{ and } Y = \frac{n(n-1)}{2c}.$$

Y, then, suggests a measure for the level of service. Obviously, for large values of n, Y can be approximated by $\frac{n^2}{2c}$, so that as demand on the system varies, the total ingression loss tends to vary with the square of the demand.

The relationship between the *marginal* ingression loss $\frac{dY}{dn}$ and the *average* per carrier loss Y/n is of interest:

(4) $$\frac{dY}{dn} = c^{-1}n, \text{ and } \frac{Y}{n} = \frac{c^{-1}n}{2}.$$

In the first place, c^{-1} can be construed as a "price" paid by each unit for the services of a transportation system at their scarcest, much in the way that we spoke of v^{-1} being a time cost of movement.[32] In the second place, the entry of the nth, or marginal, unit imposes costs on the system in the amount of nc^{-1}. If its entry into the system is randomized, over a large number of cycles the marginal unit will tend to bear the average costs in the amount of $\frac{nc^{-1}}{2}$, while an equal cost is distributed over the remaining units as a social cost. This is a scale diseconomy of the system: the entry of another unit will impose losses on the remainder of the system equal to those borne by it over a large number of cycles in which entry is completely random.

Ingression and congestion are related phemonena: given the technology of the system, ingression is the irreducible minimum of time loss resulting from the aggregation of demand. Congestion losses arise from the reduction of the "free flow," or desired, velocity imposed on a unit because of the behavior of other units in the system as exemplified in Figure 11. In a "saturated system where passing opportunities are rare or nonexistent, the movement of all units will approach a uniform flow whose velocity may be dictated by the slower speeds of the early entrants. This "system-dominant velocity" may very well be less than the "capacity velocity" of the system, and, accordingly, the system will experience a

---

[32] The round trip movement of the work trip experiences ingression loss in both directions, and by reversing the demonstration it can be shown that the ingression loss of the outflow phase is equal to that of the inflow phase. The total, round trip loss will be referred to as "cycle ingression loss."

higher level of loss than ingression analysis suggests. Congestion and ingression losses each tend to vary with the load on the system, and ingression can be construed as a measure of the congestion potential in the system. (See Appendix A.) They consitute the principal "increasing cost" element in movement systems.

The ingression losses defined are minimal under the circumstances, since all units are posited to move at maximum capacity conditions. Where units operate at less than maximum capacity, the actual amount of ingression in the system will be greater, particularly in systems with a low degree of system control. This distinction is essentially that between the traffic engineer's concepts of "base" and "practical" capacity, between ideal and real conditions of operation.

The motion diagram in Figure 10 represents three kinds of time-costs which emerge from saturated systems, such as urban traffic nets operating under peak period conditions with deadlines. The first is the "pure" time-costs DA arising from uniform motion expression:

(5) $t = v^{-1}s$, where $v = $ a "desired" velocity.

The second is pure congestion loss resulting from the impairment of desired velocity by other units sharing the system (AB or AC):

(6) $t' = (v'^{-1} - v^{-1})s$, where $v' = $ the reduced speed.

The third kind is that occasioned by the necessity of positioning oneself in the flow of units to meet deadline requirements. This loss will range from zero to EC for any individual unit, and has been referred to as pure unit ingression y so that

(7) $0 \geqslant y \geqslant \dfrac{n}{c}$, where $c = $ capacity, and
                                     $n = $ demand

The total time-cost of a round trip x_o is then described as

(8) $x_o = 2(t + t' + y)$.

Earlier it was pointed out that unit ingression y is indeterminate, with respect to any single trip. However, since the decision to enter the traffic flow is completely decentralized, it is reasonable that in the long run

(9) $\dfrac{1}{z}\displaystyle\sum_{i=1}^{z} y_i = \bar{y} = \dfrac{c^{-1}n}{2}$

where y_i = amount of ingression in ith trip
   $\bar{y}$ = average ingression for all trips
   z = number of trips
   n = current deadline demand
   c = current capacity.

That is, the average ingression experienced by any single unit over a large number of trips will approximate the average value of ingression in the system for all trips. This consideration permits the use of the average $\bar{y}$ for the indeterminate value of y.

For operational purposes, the boundary between t and t' is indeterminate because of its subjective nature. To relate time-costs to a "desired" velocity is a difficult behavioral problem, so that it is convenient to redefine v to be the *maximum permissible* velocity. Given the nature of urban traffic control, this redefinition is a step closer to the real world as well as a helpful simplification. v is identified as a policy parameter which, it is assumed, effectively binds all units—that is, no unit would freely select a velocity lower than v; t' then measures the increase in marginal time-costs (i.e., the slope of the line) resulting from a congestion induced reduction of velocity below that permissible.

Now, actually, the pure congestion losses in a system which bring about an increase in time-costs (as in the shift from OA to OB in Figure 12) can be construed as the effect of congestion imposed by a series

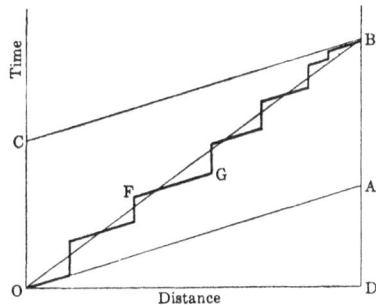

FIGURE 12. *Motion diagram.*

of capacity chokes (as in $OFGB$) whose losses accumulate even though the carrier returns to the system velocity between chokes (i.e., slope of FG = slope of OA). Congestion losses average out along the motion line OB even though the slope is always equal to OA (except, of course, for short periods of acceleration and decleration).[33] In short, the surplus

---

[33] That such an interpretation of the congestion effect is reasonable is discussed in Appendix A.

time-cost AB of a trip from O to D can be interpreted on the one hand as the result of the greater average slope OB of movement $OFGB$; on the other hand, it is more convenient for the purposes of the model to construe it wholly as an ingression effect AB deriving from the minimum capacity on the route imposed on the pure time-costs AD. Thus, if the capacity of the route is uniform over its course, the loading will result in a pure ingression loss; where capacity is variable along the route, both ingression and congestion losses will occur, but they will add up to the value of ingression computed on the point of least capacity. So long as it is understood that c always refers to the least capacity along the route, congestion losses may be absorbed into the ingression expression, and the average time-cost to the individual of a trip can be generalized:

$$(10) \qquad x_o = 2(t + \hat{y}) = 2\left(v^{-1}s + \frac{c^{-1}n}{2}\right)$$

$$(11) \quad \text{Then} \qquad x_o = \frac{n(av^2 + \gamma v + \zeta) + 2s}{v}.$$

This expression for the total time-costs x_o of a round trip between two points under peak and deadline conditions contains two independent variables,

  n = the number of demanding carriers, and
  s = effective distance, or the least distance between two points via the transportation net;

it contains several parameters:

  a = a technological parameter quantifying the deceleration characteristics of carriers in the system,
  γ = an operator reaction-time parameter relevant to decentralized systems,
  v' = a policy parameter expressing permissible velocity in the system.

ζ is a constant for the length of carriers in the system.

The conclusion emerges that the time-costs of a movement system result not only from the overcoming of distance but also from the competition of units for a limited supply of movement opportunities. And so we deal with two separate "prices" v^{-1} and c^{-1} for a good which has two economic dimensions s and n, and since the price c^{-1} is a function of v^{-1}, we can introduce fundamental technical relationships into the time-cost expression.

At the beginning of this chapter we accepted the proposition that the spatial organization of the city was a direct consequence of the manner

in which the movement of goods and persons was organized. This relationship between space and movement opportunities is subsumed in
the concept of accessibility, which has been defined as a cost concept
involving two kinds of costs: the time absorbed in movement between
points and the money-costs for the services of transportation. This
chapter has sought to identify the sources of, and factors in, the time-
costs of urban transportation by viewing the situation in the form of
a simple analogue based on some of the crucial characteristics of urban
movements. In the next chapter we will construct the second stage of
the model by examining the relationship between transportation costs
and the value of location in the urban space.

III

The Economics of the
Journey-to-Work

The Value of Time-in-Transit—Alternatives

For the worker the time consumed by the work trip is a true cost—time is a valuable commodity which must be "spent" if the trip is to be made. If this were the only cost incurred in making the work trip, the costing of the work trip would be simple indeed; that personal transportation involves other kinds of costs complicates the picture. The price of the transportation service offered by a public carrier is a fare, but the cost of a work trip in a private automobile is more diffuse: marginal trip costs would include among other items the costs of fuels and lubricants, mileage-related maintenance costs, tolls, and parking fees. Actually, personal transportation has two prices, in two different "currencies" —time and money; the effect on the economic behavior of the trip-taker is the issue to be resolved here.

There are two possible solutions to the cost problem, and because both have logical merit each will be discussed, so that the reasons for preferring one will be clear.[1] The two solutions arise over the issue of the "exchangeability" of one currency for another. If exchangeability is permitted, the problem is subject to the conventional market solution— the consumer adjusting his consumption of a particular good (here the work trip) to its cost to him when all other prices are held constant. Where such an exchange is not possible, the market solution is much more complex, the conditions for consumer equilibrium are different, and in aggregate behavior there is considerable latitude for ambiguity.

The exchangeability assumption grows out of classical labor market

[1] I am grateful to David Milstein for numerous discussions on this problem. Milstein's research on consumer behavior in outdoor recreation provided fruitful insights which have been incorporated into the following discussion.

theory, which posits a competitive market for the worker's time. In such a market the time he spends in activities other than his work has direct and assessable opportunity costs, for in reality such activities cost him the wages he could have alternatively earned. Thus, under the assumptions of a perfectly competitive labor market, time-costs can be valued in money terms and added with direct costs into a single work trip "price."

The position can be maintained, however, that the real world labor market is so imperfect a mechanism for valuing the worker's time at the margin that it does not provide the conditions of exchangeability necessary to measure the opportunity costs of other than socially productive uses of his time. The conventional market conditions for bargaining between free sellers and purchasers of labor services are abridged by the rigorous organization of both sides of the market; not only has price-making in the labor market become partly political, but the manner and degree in which the worker may sell his services have become narrowly circumscribed by the institutional organization of production. In addition, there are severe immobilities confronting the worker which impair his adjustment to changing economic circumstances. Finally, time is not a homogeneous commodity as is currency—an hour of time does not have the same value for the individual at 3 a.m. that it has at 3 p.m., simply because of the difference in alternative possibilities for using it. All of these points argue against the assumption that the wage rate represents the marginal value of a worker's time *to himself,* and if they are conclusive, we must discard the description of individual behavior in the allocation of personal time in classical economic theory.

Individual behavior under "double currency" conditions has its analogy in the point rationing problem that was discussed by Dorfman, Samuelson, and Solow.[2] If one has a budget of ration points (or time, in our case) and one of money, and if there are a number of commodities which have prices in ration points and in money, then with respect to any pair of such commodities there will be two possible consumption outcomes depending on the relative prices and the relative utilities of the commodities (see Figure 13). Either ration points or money will be binding and the other redundant.[3] The binding "cur-

[2] Robert Dorfman, Paul A. Samuelson, and Robert M. Solow, *Linear Programming and Economic Analysis* (New York: McGraw-Hill, 1958), pp. 25-27.

[3] The exhaustion of both is merely the boundary condition between the two outcomes and can be interpreted to mean that either one or the other currency is binding.

rency" will always be the scarcer, the one whose alternative consumption possibilities are more highly valued by the individual.

When we relate this concept to the manner in which individuals make choices about transportation (and ultimately about household location), it is apparent that in the absence of exchangeability a worker may be behaving as though the scarcity either of his time or his money were governing his behavior, and in either case the other is ineffectual in

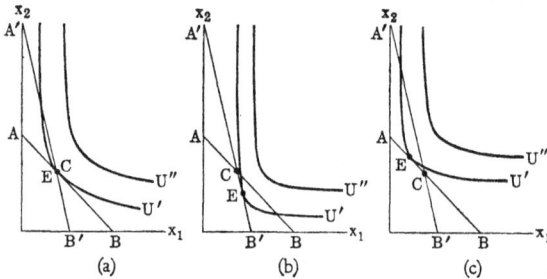

(a)　　　　　　　(b)　　　　　　　(c)

FIGURE 13. *The lines AB and A'B' represent the dollar and ration-point budget equations, respectively. The heavy locus ACB' represents the locus available to the consumer since the "scarcest currency" is always the bottleneck. In a, this locus touches but does not cross the highest indifference curve at C. In b, this phenomenon occurs along CB', where dollars are redundant; in c, the ration points are redundant. When there are only two x's and when both constraints are known to be binding, there is no room left for maximizing behavior; only when there are more goods does the problem become interesting.*

Source: Robert Dorfman, Paul A. Samuelson, and Robert M. Solow, *Linear Programming and Economic Analysis,* The RAND Series (New York: McGraw-Hill, 1958), p. 26, Figs. 2-4.

allocation. This formulation leads to several complications in the costing of transportation. First, in a population of trip-takers the total pattern of transportation behavior may be very complex—for some of the trip-takers time may be the binding constraint, for others money. Second, autonomous changes in critical utility and price parameters may lead to changes difficult to predict. If the money budget is binding, a fall in money prices will have conventional effects only to that point where suddenly time becomes the limiting constraint, and further decline in money prices will have no direct consumption effect.[4] Similar unpredictable effects might result from a change in the total money budget, from a reduction in time-costs, or from shifts in the preferences of the trip-taker.

These conclusions suggest further that the constraints may be distributed among the population in accordance with income levels. Money is more likely to be the binding constraint for low income groups, time for upper income groups. Responses of the two groups to parametric change would be quite different, so that groups whose incomes were secularly rising would experience a change in the binding constraint at some point. Further, the central clustering of low income groups would be partly explained by the attempt to minimize money costs of transportation, the suburban movement of upper income groups by their willingness to purchase automobiles to compensate for the time losses of their greater removal from employment sites.

However, to argue that the market mechanism for converting human time into an analogous money value is imperfect is not to say that it is inoperable. Granting that the real world labor market is considerably removed from that posited by economic theory, nevertheless it does directly function to set a price upon productive time purchased by the employer as an input into the processes of production. In fact, we do not have to argue that this price is relevant to other allocations of the worker's time at all; we have only to illustrate the mechanism by which the time spent in the journey-to-work enters into the market processes establishing the wage rate, or the price of the worker's time, in order to support a single currency formulation of the time costs of transportation.

The Marginal Value of Leisure—
An Approach to the Valuation of Time

We take it for granted that the worker has a reasonable awareness of the array of employment (income) possibilities open to him, and that he responds to the satisfactions inherent in the opportunities for spending his leisure. We also assume rationality in the behavior of the worker, so that each succeeding employment decision will be made to increase his total satisfaction and so that at any point in time his com-

---

[4] In the sense that more transportation service would be consumed. For goods not involving time costs such a decline in the money costs of transportation would have the effect of freeing income so that it could be allocated to them.

bination of leisure and employment approximates a short-run optimum for him. His ability to adjust at any moment may be limited by his immobilities, but we assume that when he does adjust he makes the best decision available to him.

Hence, our argument is based on the simple observation that the market behavior of the individual worker can be described as a choice among combinations of income and leisure, both of which he values positively;[5] given two alternatives having the same income (and other advantages) but differing in the quantity of leisure remaining to him, he will choose that affording the greatest amount of leisure, and vice versa. For any such combination the important relationship is the rate at which he will substitute income for leisure to maintain the level of total satisfaction, that is, "the marginal value of leisure."[6] This relationship can be restated, in terms of the worker's response to the prices offered for his services in the market, as a curve expressing the wage rate necessary to induce him to give up one additional hour of leisure daily, which will be referred to as the *marginal-value-of-leisure* (or MVL) curve.[7] (Figure 14). This curve has a positive slope reflecting

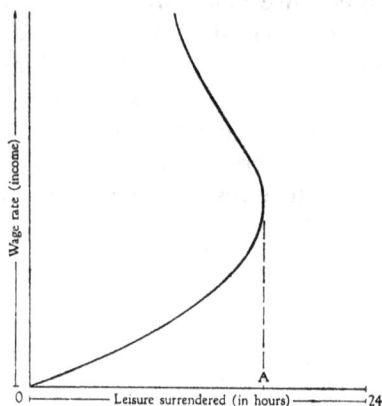

FIGURE 14. *Marginal-value-of-leisure-curve (schematic).*

[5] The following exposition leans heavily on the theoretical discussion of the labor market in the following sources: Tibor Scitovsky, *Welfare and Competition* (New York: Richard D. Irwin, 1951), pp. 83-90; Kenneth E. Boulding, *Economic Analysis* (rev. ed.), (New York: Harper & Bros., 1948), pp. 220-33, 742-46.

[6] As defined by Scitovsky, *op. cit.,* p. 86.

[7] This conclusion leans heavily on the assumption of full employment in the labor market. Where less than full employment obtains, the firm will tend to shift costs back to the worker through the mechanism of competition among

the condition that time is valuable to the worker, and it is concave upward to reflect the principle of diminishing marginal utility—the more
leisure one has, the less valued is the last hour of it, and, conversely,
there is some point (*A* in Figure 14) at which the marginal hour will
be so highly valued that it cannot be bought.

In a perfectly competitive economy, the primary condition for the
optimum allocation of labor among production possibilities is the equivalence of the worker's marginal value of leisure with the marginal cost
of labor to the firm, or the wage rate. The fact that the labor market is
imperfectly competitive, however, as has been acknowledged above,
vacates any logical assurance that this equivalence holds. Instead, the
argument is thrown back upon an inequality: the wage rate for which
labor services are supplied is equal to *or greater than* the worker's
marginal value of leisure under conditions of full employment.[8] The
gap between the wage rate and the MVL curve can be construed as a
measure of the imperfection of the market; or, otherwise expressed, the
degree to which competitive market conditions obtain will determine
the extent of the departure of wages from the marginal conditions. The
following discussion is based on the equivalence of the marginal value
of leisure and the wage rate as the limiting condition, and the implications of imperfect market conditions will be suggested where relevant.

Implicit in conventional labor market theory is the assumption that
the amount of a worker's input into the production processes is equal to
the amount of leisure time he gives up. In fact, however, this equivalence
does not hold; the worker gives up more leisure time than he puts into
work, and the difference is the time spent in the journey-to-work, the
"time-costs" previously discussed. In Figure 15, *OM* is a worker's
marginal-value-of-leisure curve; a wage of *PH* would induce him to
surrender an amount of leisure *OP*, and a wage of *QJ* an amount of
leisure *OQ*. Since every point on *OM* represents a unique combination
of wages and leisure time surrendered, *aa, bb, cc,* etc., a set of daily
"iso-income" lines, are each the locus of hours and wage rates yielding

---

workers in the labor market, depressing the marginal-value-of-leisure (or "price
offer") curve. In this sense, the disposition of the curve is dependent on employment conditions in the short run, or where there is a high degree of immobility of
labor among the labor markets. For the remainder of this discussion, it will be
understood that the full employment assumption is relevant.

    [8] Scitovsky labels this the "price-offer curve," and Boulding the "individual labor
supply schedule." These are identical with the expression "marginal-value-of-
leisure curve" as used here.

the same daily income. Now assume that the employing industry is concentrated at a single point and the worker's domicile is so located that
his journey to and from work consumes a period of time equal to PQ.
Assume further that the technical and institutional character of the
industry's processes of production requires the input of a P-hour working
day, no more, no less. To work a P-hour day, the worker must give up
Q hours of leisure, necessitating a wage rate of QJ for a total of Q hours.
QJ falls on the iso-income line $a'a'$, which is the daily wage necessary

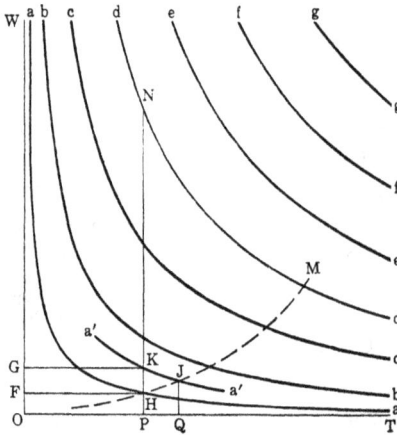

FIGURE 15.

to induce the worker to surrender Q hours. The industry, however,
pays hourly wages for only P hours of work per day; and hence must
pay an hourly wage to meet the daily wage $a'a'$ which is determined by
the intersection at K of the indicated iso-income line $a'a'$ and the length-
of-workday line PN. PK, then, is the "manifest" hourly wage rate—the
price the firm must pay for the specified labor inputs. If the journey-
to-work took no time at all, so that the amount of the labor input would
be equal to the amount of leisure surrendered, the worker would be
willing to work a P-hour day for a wage of PH. This is the "pure"
hourly wage rate. The difference between his manifest and pure hourly
wage rates is the simple consequence of the worker's home-work separation. The value of the daily time-cost of this separation is the difference
between the manifest and pure *daily* wage rates, indicated by the area
$FGKH$ in Figure 15.

(12) $$P_o(X_o) = W - W'$$

where $P_o(X_o)$ = value of time spent in the journey-to-work
W = manifest *daily* wage rate ($OGKP$ in Figure 15)
W' = pure *daily* wage rate ($OFHP$ in Figure 15).

Now if the daily wage rate is equal to the product of the hours worked and the hourly wage rate, and if the MVL curve is defined as $p_o(\mu)$, the value of the time-costs of the journey-to-work x_o is determined by the MVL function and the amount of time consumed daily by the work trip:

(13) $$p_o(x_o) = p_o(\mu + x_o) - p_o(\mu)$$

where $x_o = x_o(s, v, n, c)$
and μ = the length of the working day, an institutional constant
x_o = the time spent in the journey to and from work
p_o = "a marginal-value-of-leisure function of."

Given the fact that industry purchases labor at the manifest hourly wage rate, the inference arises that the time-costs of the journey-to-work enter the production costs of the local firms as a component of their labor costs.[9] The simplest empirical evidence to back up this inference is the experience of firms which have removed to locations peripheral to large urban areas but remote from the labor force and have been compelled to offer premium wages to induce certain types of labor (especially female clerical help) to make the extended work trip. In short, the costs of the journey-to-work enter into the costs of production and are borne by the consumer, and may have a significant impact upon a firm's relative market advantage.

The "two currency" formulation of the costs of the journey-to-work is not persuasive when the work trip is viewed not as an economic good in itself but as an offset against the returns from employment. Since the worker's decision is based on the *net* returns to him of employment, his ability to shift his transportation costs forward to the employer provides directly for the exchangeability of time and money currencies. Application of the two-currency formulation must be confined to those situations in which this shifting is not feasible—where there is under-

---

[9] This contradicts a conclusion reached by Leo Schnore: ". . . the immediate costs of commuting—in time, money, and energy—are borne by the individual worker, and . . . the money costs are well hidden in the family budget." Leo Schnore, "The Journey to Work in 1975," in Donald J. Bogue (ed.), *Applications of Demography: The Population Situation in the U.S. in 1975* (Scripps Foundation, Miami University, 1957), p. 75.

employment of labor, monopsony in the labor market, or a highly elastic demand curve for labor, and, in each of these cases, only where there is a substantial geographical, industrial, or occupational immobility of labor. The two-currency relationship is thus a short-run phenomenon as well, for time tends to level the obstacles to labor mobility, permitting the laborer to escape from circumstances that restrict him from achieving his optimum labor market position.

If we acknowledge (1) the qualifications of the exchangeability condition, and (2) imperfections of the labor market, the conventional economic concept of the labor market affords a logically consistent mechanism by which the time-costs of the journey-to-work can be valued in money terms. As was previously suggested, we can expect most combinations of wage rates and leisure time given up to fall above the MVL curve, but this fact need trouble us no more than the existence of positive profits disturbs the theory of the firm. The wage rate provides a measure of the objective value to society of the worker's time, and hence can be used in the exchange. If he values time little, he will view the surplus as a windfall, and a decline in daily wages will not affect his behavior until the windfall is exhausted. The important effect will show up elsewhere—in the rate at which he is willing to convert time savings into payments for locational advantages. This situation will be taken up at a later point.

Transportation Costs in the Journey-to-Work

In addition to the time-costs, direct outlays by the worker for transportation must be brought into the calculations, and the impact of these will vary with the mode of transportation. However valued, they are likewise costs chargeable against the gross returns from the worker's employment and can be treated in a fashion similar to that of the time-costs. They can be expressed as a constant with respect to time and distance (within limits), as in a single-fare transit system; a function of distance, as in auto mileage costs; or a function of time, as may be expended by automobiles in highly congested traffic systems.

For the purposes of putting together a model based on the costs of the separation of home and work, we are interested basically in two classes of money costs in transportation—those costs which vary with the distance traveled and those which vary with the number of trips taken. This permits us to exclude the complexities involved in the allocation of capital and overhead costs associated with private automobile ownership, since their dimensions in the family budget would not affect separation costs unless the automobile were indispensable as the exclusive mode of travel for the work trip and were used for the work trip exclusively. These cases are excluded for the sake of simplicity, and we will confine our attention to distance-variable inputs and trip-variable inputs as the sole components of household transportation costs associated with home-work separation.

Thus, the daily money costs of the work trip can be expressed in the following form:

$$(14) \qquad \sum_a p_a x_a + \sum_b p_b x_b$$

where x_a = the amount of distance-variable input "a" consumed, that is, $x_a = x_a(s_i)$ where s_i is the effective distance separating home and employment site

p_a = the unit cost of "a"

p_b = the cost of trip-variable input "b"

x_b = the number of daily trips.

Now since x_a and x_o are both functions of s_i, it follows that x_a can be expressed as a function of x_o, and for all positive values of x_o a new curve segment is generated ($H'J'M'$ in Figure 16 in place of the corresponding segment HJM of the MVL curve) which will leave the worker in the same net satisfaction position he would have enjoyed on HJM if transportation had been free. The manifest *daily* wage is raised from iso-income curve $a'a'$ to $a''a''$ because of QJ' intersecting $H'J'M'$ at J', yielding a manifest *hourly* wage rate of G'. The increment to the manifest hourly wage rate effected by the money costs of transportation is

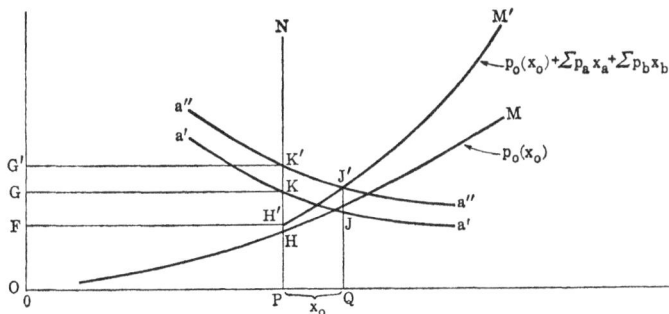

FIGURE 16.

GG', where the total daily cost is represented by $GG'K'K$. The process whereby the time-costs of the work trip are bid into the firm's production costs is equally applicable to these money costs, and hence they, too, can be calculated as production costs.

The value of the time-costs as well as the out-of-pocket costs having been identified, it is now possible to express the total transportation costs associated with spatial separation of the worker's residence from the place of his employment:

(15) $$X = p_o(x_o) + \sum_a p_a x_a + \sum_b p_b x_b.$$

Since X represents the cost per working day, it must be summed up over the number of working days per annum to yield an annual cost figure (we assume sufficient stability of the MVL curve and prices to permit the multiplication of X by the number of working days per year). These are, of course, private costs. They do not reflect subsidies or social costs of any sort, for here we are concerned exclusively with the economic behavior of the individual. The important feature of this expression to keep in mind is that both x_a and x_o are functions of the length of the journey-to-work and the demand upon the system by which the journey-to-work is made; we can abbreviate as follows:

(16) $$X_{i-j} = \phi(s_{i-j}, n_j, p_a, x_a, p_b, x_b)$$

where n_j = deadline demand on system
  s_{i-j} = distance on system between the worker's domicile at i and his employment site at j, and
  ϕ = "a cost function of"
All other parameters and variables are held constant.

This formulation can relate to any mode of transportation employed in the work trip; it will hereafter be referred to as the "transportation cost function" of home-work separation.

IV

Economic Dimensions of Urban Space

Position Rent

Once the worker's transportation cost function is established, the transportation technology described, and the number and behavior of other users characterized, the economic qualities of the space in which the worker functions become defined. Thus, for any employment site j and any residence site i the annual costs to worker k of the journey-to-work $_kX_{i-j}$ can be ascertained.[1] Following the postulate that each worker seeks to maximize his net income, and assuming that the wage rate for all k and all j is constant and uniform, he will seek the "least-cost" combination of i and j, and here we can distinguish two separate cases: (1) every worker k finds an optimum position at a combination of i and j at which no other worker finds his least-cost position, or (2) at least one worker finds his optimum position at an i-j combination which is also the best position for at least one other. In the first case, all k locate at their optimum positions on a one-to-one matching basis. The second case presents a problem in the allocation of space.

Equilibrium in economic analysis requires that the returns to labor be equal to the worker's marginal value of leisure. If we assume the marginal-value-of leisure function to be uniform for all workers, the returns to labor must hence be equal for all workers. However, this can be true only in terms of labor's *gross* returns as measured by the manifest hourly wage rate. The *net* returns can be defined as the manifest daily wage *less* the daily costs of the journey-to-work. Since the latter vary with the location of the residence i with respect to any employment site j, the net returns to all workers cannot be equal (except in the special instance of case 1 above). The worker for whom X_{i-j} is

---

[1] All X_{i-j} such that $_kX_{i-j} + _kW' < _kW_i$ are excluded by our assumption that wages must be at least as great as the marginal value of leisure. (See Equation 12.)

very small—who lives next to the plant, for example—thus receives a surplus over and above the marginal value of his leisure, and since X_{i-j} is a function in part of the distance between employment site and residence, each worker would receive such a premium in varying degree as he lived near his employment center, except for the worker domiciled at the marginal location m (Figure 17). This premium results not from his reluctance to accept employment at a lower wage rate but from the situational advantage associated with the location of his residence vis-à-vis the employment center.

Some residence locations thus have an economic advantage for the worker, and he will compete with all others to capture the advantage. The "price" of any individual location will be bid up to the point where all workers but one are excluded,[2] with the result that part or all of the value of the locational advantage is absorbed.[3] Thus, a unique set of location rents is generated for every worker for every point in space. These play a role in equalizing the net returns among all members of the labor force and at the same time provide a structure in space of urban site values. This form of location rent—the annual savings in transportation costs compared to the highest cost location in use—will be referred to as "position rent" R.

Several things emerge from this formulation. First, the manifest daily wage is set by the worker at the margin of the market, m in Figure 17.

[2] "This has two implications: First, with prevailing price relations the individual finds his highest utility *here*. Secondly, at the price paid for this site he *alone* finds his highest utility here." August Lösch, *The Economics of Location* (New Haven: Yale University Press, 1954), p. 247.

[3] Alonso seems to argue that the total value of the advantage will be absorbed by rent (*A Model of the Urban Land Market*, p. 138), although he cautions elsewhere (pp. 143-45) against the confusion growing out of the several senses in which the term "rent" is used. If we define rent to be the value of the site in its most productive use, then the statement is tautological and the issue is whether or not the owner of the land can capture all of the rent from the entrepreneur. It would seem that this is possible only under the restrictive assumptions of perfect competition and where all bidders are of the same industry (uniform technology) and equally efficient. We can move to a more general statement by relaxing the perfect competition assumptions and viewing rent as the value of the joint product of site and entrepreneurial capacity, so that the joint return should be allocated between them in proportion to the incremental productivities of each at the given site over their next most productive employments, but under any circumstances the payments to the entrepreneur and to the land owner will each be greater than could have been earned in any alternate employment. This sets a range within which the entrepreneur and the land owner will share the rent, depending on the bargaining skills of both.

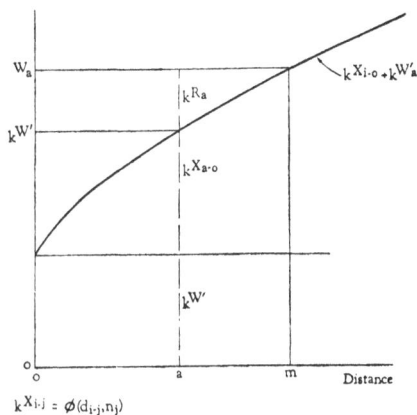

FIGURE 17.

$_kX_{i \cdot j} = \phi(d_{i \cdot j}, n_j)$

Second, the manifest daily wage includes not only pure wages and the transportation costs of the worker but also his position rent. Third, both transportation costs and the position rents of the workers enter into the costs of production, so that their incidence falls upon the consumer of local production.

The conditions for an equilibrium distribution of worker's residences can be described as resulting from the effects of position rent under the circumstances specified above. Assume that there are h points, or positions, at which a set of g ($g < h$) households may locate with respect to a single employment center at o. These points may be expressed in terms of their distance from the center s_{i-o}, so that each is associated with a unique $_kX_{i-o}$. Then for any subset of two points an incremental rent may be charged for the nearer one equal to the saving in transportation cost associated with occupying it rather than the other. So in Figure 17, for points a and m, $_kR_a$ may be charged, and since the sum of transport costs and position rents is equal for both point a and point m, neither offers an economic advantage over the other—they are, hence, locationally indifferent. This same condition can be extended to h locations and h-1 households, so that there is at least one set of position rents which will make location a matter of indifference to the households. This is an equilibrium set of rents and may be defined as

$$(17) \qquad _kR_i + {_kX_{i-o}} = L, \quad \begin{aligned} k &= 1, 2, 3, \ldots g \\ i &= 1, 2, 3, \ldots \ldots h \end{aligned}$$

where $_kR_i$ = the position rent for worker k at any point i
  $_kX_{i-o}$ = the annual transportation costs to worker k of the journey-to-work from any point i to the employment center at o
  L = location costs, a constant for all i and k.

Where the number of units demanding positions is less than the number of positions available, the units will successively occupy positions moving out from the center until the units are exhausted, and L will be set by the $_kX_{i-o}$ associated with the nearest unoccupied position (for the last position occupied would charge a rent equal to the transportation saving over the first unoccupied position), whose position rent is zero. This is the "margin point," whose transportation costs $_kX_{m-o} = L$. Where the number of units demanding positions is greater than the number of positions available, the rent increment *between any two positions* will be identical with that of the previous case, but there will be a tendency to bid the entire structure of rents up to some point where the value of L will drive the surplus of demanding units from the market. Let this value be L' at the equilibrium point. Three conditions can be identified, where the number of positions is b and the number of units is a:

$$\text{if } a < b, \text{ then } L = {}_kX_{m-o}$$
$$\text{if } a > b, \text{ then } L = L'$$
$$\text{if } a = b, \text{ then } {}_kX_{m-o} < L < L'.$$

The aggregate level of position rents is determined by demand-supply relationships, while the incremental structure of rents in space is a function of transport costs[4] (see Figure 18). In other words, transport costs will determine the relationship that must exist among any equilibrium set of position rents, while the "market" will select the equilibrium set.

This discussion of position rents reveals the effect of transportation costs on the structure of incremental rents under two restrictive assumptions: (1) for any individual unit the demand for space is completely inelastic, and (2) there is a finite number of discrete positioning opportunities available. Although these assumptions do not represent the real dimensions of the market for space, they may have relevance in part: the inelasticity assumption may very well be a good representa-

---

[4] Benjamin Stevens distinguishes between *absolute* and *differential* rent, the former measuring the scarcity value and the latter quality differences, where "quality" includes locational advantages. In the formulation above,

$$\text{absolute rent} = L - {}_kX_{m-o}, \text{ where } L \leqslant {}_kX_{m-o}$$
$$\text{differential rent at } i = {}_kX_{m-o} - {}_kX_{i-o}$$

Benjamin H. Stevens, "An Interregional Linear Programming Model," *Journal of Regional Science*, Summer 1958, Vol. 1, No. 1, pp. 60-98 (see especially pp. 84-86).

tion of the demand for office space by certain classes of administrative offices, and the limited supply opportunity assumption is perhaps not unrealistic for specialized real estate. The main purpose has been to demonstrate the manner in which the "frictions of space" as measured by transportation costs affects the distribution of urban rents in space. If unit demand for space were completely inelastic, the structure of location rents should be a good representation of the real world. In the succeeding discussions we will relax these assumptions, assuming instead (1) that the individual demand for space is generally elastic, and (2) that the supply of space is continuous and unlimited.

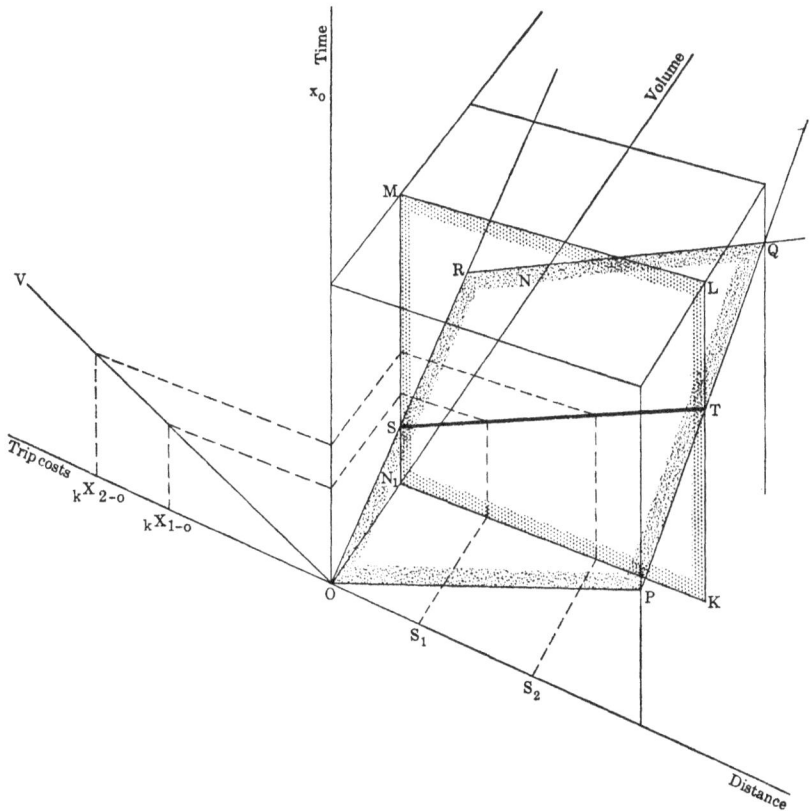

FIGURE 18. *Simultaneous determination of incremental position rents* R.
Plane $\overline{OPQR}$ = *transportation function* = $v^{-1} d + c^{-1} N$
Curve $\overline{OV}$ = *modified marginal-value-of-leisure function* = $p_o (x_o)$
R *at* s_1 *with respect to* s_2 = $_kX_{2-0} - {}_kX_{1-0}$.

Given the transportation function expressed as a function of distance from the center s_o it becomes possible to express the spatial structure of position rents by rephrasing Equation 17, above.

(18) $$_kR_i = L - {_kX_{i-o}} = {_kX_{m-o}} - {_kX_{i-o}}$$

where L = location cost
  $_kR_i$ = position rent at a point i
  $_kX_{i-o}$ = transportation costs between center and point i for household k
  $_kX_{m-o}$ = transportation costs to the margin point m from the center for household k.

Figure 19 is a schematic expression of the above relationship: a household located at i enjoys a premium in transportation costs with respect to a household located at the margin m. This premium invites competition from all households located at a greater distance for the position at i, since the last household can offer a rent for the position at i equal to the difference in the transportation costs. In Figure 19 this rent would be R_i. In this fashion a locational equilibrium is established by a structure of rents which would absorb the locational premiums. No household could increase its net returns by changing location, and no location could increase its returns by changing occupants. In a purely static

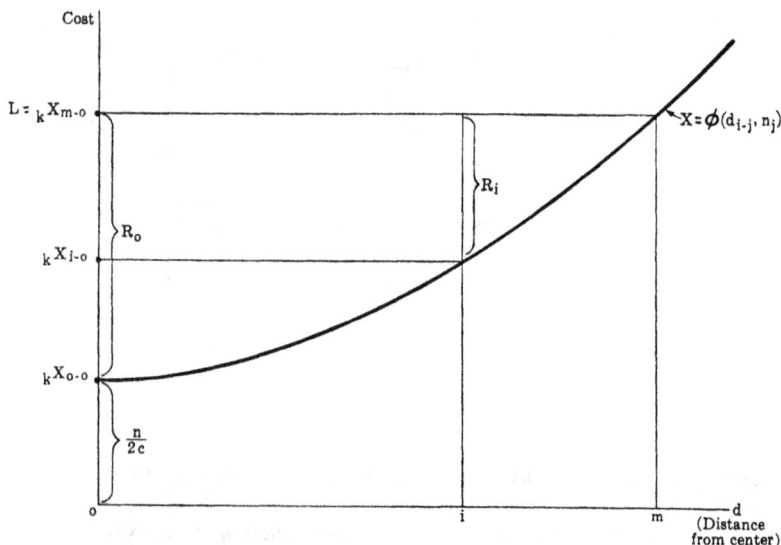

FIGURE 19. *The spatial structure of position rents.*

sense, this is the spatial optimum toward which a perfectly competitive urban economy would move in the distribution of households.

On the Empirical Determination of Position Rents

The definition of position rent offers a strategic point to stop and review what has been brought forward in the construction of the model up to this point, since position rent is the central parameter in the model. For the first model stages position rent is the effect of an array of variables arising from the technology of transportation on the one hand and from the structure of the local economy on the other. For the final stages of the model it is the "decision variable" by which the economic household will make its dual decision of how much space to consume and where.

Figure 20 summarizes the relationship of position rent to the input variables and intermediate variables discussed so far. Note that position rent $_kR_{i-j}$ is not determinate as far as we have pursued it, since it requires a feedback loop ($_kR_{i-j}$ to [$_kq_i$, $_kr_i$] to $_kX_{m-o}$ and back to $_kR_{i-j}$) to establish equilibrium conditions. In short, a precondition for the establishment of position rent is determination of the margin point m, which depends on information about the densities at which the population is distributed—information which is outside of the model at this point.

From this schema we can identify the two major classes of information required by the model (Boxes 1 and 2 in Figure 20). The first consists of certain kinds of information about the spatial setting of the model (the "i-j" relationships), and the second consists of information about the population and its relevant characteristics (the elements indicated by N, n_j, and variables with the subscript k). Finally the policy variable v must be identified, and data on the money outlays for transportation will have to be developed. This slate of information will permit determination of the transportation cost surface $_kX_{i-j}$, that is, the annual cost of transportation for any individual k (out of a population N) between any domicile site i and any employment site j. To carry out this determination several specialized functions are hooked together: a capacity function, an ingression function, a marginal-value-of-leisure function, and a space demand function.

The importance of the i-j information is that it sets up the technical conditions for the model. The first step is to characterize i and j, that is,

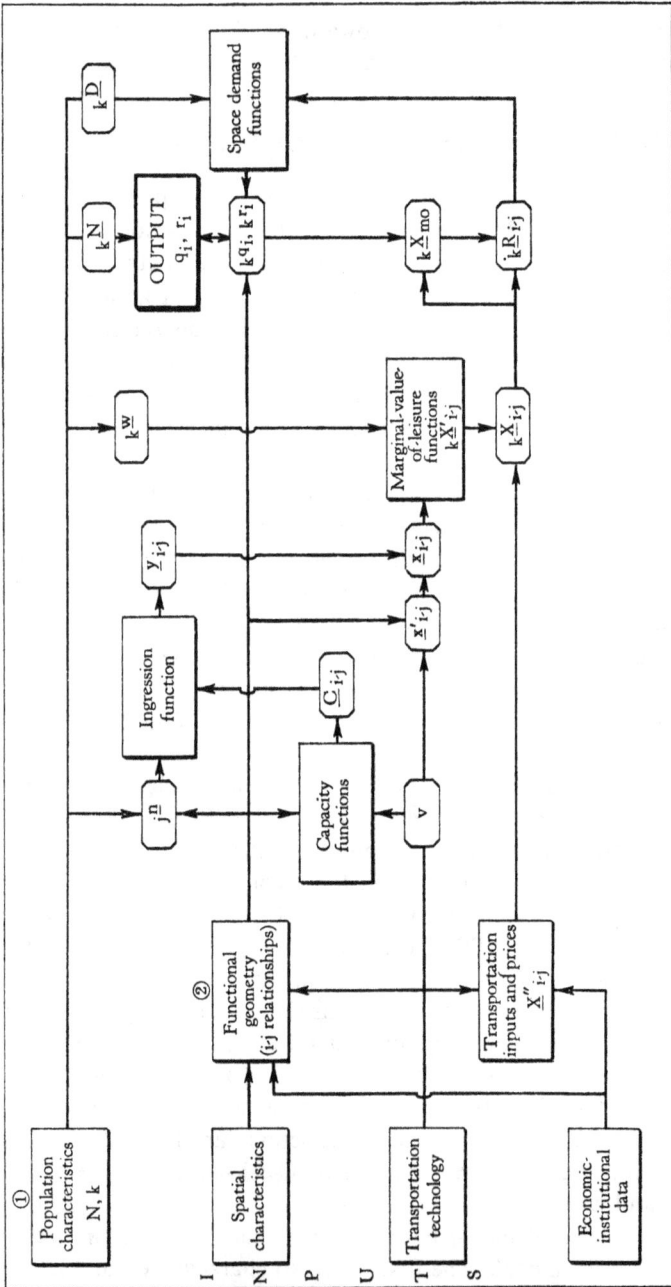

FIGURE 20. *Information flow diagram.*

to make explicit our assumptions about the distribution of employment sites j and about what parts of the space are available for household occupancy. Since we expect the model to tell us something about the distribution of population with respect to employment centers, these centers can be viewed as fixed in number and location, while i can be viewed as the general spatial variable. (A secondary use of the model might take i as the fixed element and seek information about some optimum location of employment centers.) To simplify further the presentation, it will generally be assumed that there is but one employment site o, where all employing firms are located, and thus our main interest will be focused on the gradient of densities and land values about this point, although in actual implementation such simplification could not be justified.

Once i-j is anchored by definition of j, three types of information are required: on the one hand, the employment component of the population must be distributed among the employment sites (n_j) to establish the level of transportation demand; on the other hand, what might be called the transportation technology of the i-j relationships must be described by determination of (1) the *effective distance* via the transportation net between all j and all i, and (2) the *capacity* of the transportation net between all i and all j. It is, of course, possible that several modes of transportation may be available between some i and some j. In these cases, it is assumed that the mode offering movement at the lowest total cost sets the cost of transportation and that the surplus costs of more expensive forms of transportation are actually payments for amenities associated with, but not part of, the transportation movement. Finally, by identifying the constant v (the constrained velocity) we are in position to determine the total time costs $x_{i\text{-}j}$, which are common for all workers, given the transportation technology and the spatial distribution of employment.

A comment on the production and presentation of the $x_{i\text{-}j}$ data is in order. Here there are choices to be made that are germane to the implementation of the model as a whole. In terms of the development of the data one can choose between computational and empirical values for $x_{i\text{-}j}$. Such data are frequently developed directly from auto traffic surveys by use of "floating car" techniques modified by allowances for terminal time to approximate portal-to-portal time expenditures.[5] When

[5] "One reason for the apparent variation in the results of the intra- and interurban investigations is that these studies did not include terminal time of terminal

these empirical values are mapped directly onto geography (Figure 11), they confront us with two serious limitations. In the first place, the spatial irregularities of the isochrons[6] are the aggregated effects of a variety of factors, such as geographic barriers, variations in transportation facilities, differences in demand, and the distorting effects of large institutional holdings; they present us with descriptive rather than analytical information. In the second place, empirical values afford us little or no basis for predicting future values. Computational inputs, on the other hand, are limited by the simple fact that they are surrogates for reality and tend to diverge from it in projections because they focus on certain underlying variables and exclude the rest. These limitations suggest that a satisfactory technique might involve the "splicing" of computed delay values to empirically determined ones. In this way we could cleave close to a current reality and preserve the advantages for projection of the theoretical functions.

A format is required, too, to put the data into the most useful perspective. Since the model deals with spatial constructs of one form or another, a geometrical representation is suggested, although applied models have tended to favor "cellular" forms of presentation. The value of the geometric forms lies in their ability to treat the variables as continuous over the total urban space and to describe an integral spatial structure of the variable so that i is a point in space for which specific values of the variables can be identified. In the cellular representation, space is transformed into a set of information cells, each of which has a common schedule of information but differs from the others in the empirical values assigned to the schedules. In this format the i-j framework of our model can be viewed as a series of matrices, each conveying one set of relationships between the employment sites j and the given number of residential sites i. Although spatial structure is implicit in these separate boxes of information, it is not emphasized and is not communicated clearly. Both formats are useful and say the same thing, but in different language and for different purposes: the cellular

---

effect in the measurement of distance. . . . In intraurban travel where the median travel time is usually less than 20 minutes, a 5 to 6 minute terminal time would have considerable effect on total trip time." Hansen, *op. cit.* (Ch. I, fn. 13), p. 75. The use of the "floating car" technique for collecting travel time data is described in E. M. Hall and Stephen George, Jr., "Travel Time—An Effective Measure of Congestion and Level of Service," Highway Research Board, *Proceedings 1959*, pp. 513-14.

[6] Isochron—the locus of all points in space requiring a specified travel time to reach from a specified origin.

format is most useful in the handling of data, the geometric in the communication of structure in space. For the purposes of this study the geometric expression is used for its communication value only.

Returning now to the problem of empirical implementation, it is necessary to bring in another dimension by developing data on certain characteristics of the population: income data in the form of imputed hourly wages $_kw$, and the institutional pattern of work as expressed in the length of the work day $_k\mu$ and the frequency of work periods M.[7] These k inputs and the objective data x_{i-j} implement the MVL function in the valuing of the time-costs of the separation between home and work.

Reliance upon the marginal-value-of-leisure function confronts us again with the consumer preference problem; the function is subjective and hence difficult to endow with objective qualities necessary for empirical research. One must simply assume the important characteristics of this function. For any individual, once the amount of leisure surrendered, transportation outlays, and the hourly wage rate are established, we can assume that the net daily wage coordinates lie on the individual's MVL curve. If the principle of diminishing marginal utility applies to leisure, this curve would be nonlinear and the degree of curvature would be difficult to establish over any extensive range of the curve. However, in the vicinity of the point established by the manifest hourly wage rate and the total leisure surrendered, the deviation of the real curve from a hypothesized linear function passing through this point and the origin would be comparatively small, and so for empirical purposes the assumption of linearity should not lead us too far astray as long as three aspects of it are kept in mind: (1) it will lead to a consistent understatement of the value of time-costs to the individual; (2) the degree of this understatement will be directly related to the degree to which the values for the time variable depart from the original point; and (3) such hypothesized MVL functions probably have little relevance to very large incomes. With these conditions in mind, we can approximate the value of the time-costs of the journey-to-work as follows:

For any worker A, given the following data:

$_aw$ = the manifest hourly wage rate for A
$_aw'$ = the pure hourly wage rate for A
$_ax_o$ = the daily time spent in the work trip by A
$_aX$ = daily value of time spent in the work trip by A
$_aX'$ = the daily money costs of the work trip to A
and μ = the length of the work day

[7] Certain other population data are ignored at this point.

and assuming that the MVL curve
> (1) is linear, and
> (2) passes through the wage-time point of A,

then it can be shown that

$$_aw' = \frac{_aw\mu - _aX'}{\mu + _ax_o}$$

and

$$_aX = _aw\mu - \left(\frac{\mu}{\mu + _ax_o}\right)^2 (_aw\mu - _aX')$$

then for any other value of x_o, say $x_o^*(_ax_o^* \neq _ax_o)$, a new cost X^* will be defined along the posited MVL curve so that

$$X^* = \frac{(_aw\mu - _aX')}{(\mu + _ax_o)^2} (2\mu_ax_o^* + _ax_o^{*2}) + _aX'.$$

The total scope of the computations can be held down to a comparatively small number of MVL relationships by the breaking down of the $_kw$ and $_k\mu$ into a small number of large classes. For each combination of classes of w and μ an archetypal MVL function can be defined. For the purposes of this demonstration, it will be assumed that there is a single income class and hence a single MVL function, an artifice whose purpose is to simplify only.

Another set of data required is that relating to the actual money costs of transportation—transit fares, fuel costs, tolls, parking costs, etc.—including specific inputs and their prices. We are not concerned here with the impact of subsidies unless they actually are variable and tend to have a differential effect on behavior of households over a period of time.

These data will permit us to establish the transportation cost function for any household k with respect to any given employment site j for any point in the urban region i. Once this $_kX_{i-j}$ is defined, position rent can be defined as a function of transportation costs at the margin m:

(19) $$_kR_i = m(_kX_{i-j}).$$

The Spatial Framework

Spatial models of economic activities generally begin with certain simplifying assumptions about space and rarely provide for any effective application of the conclusions of the model to real world spatial relationships. These tend to be quite complex because of innate geographical features and because of the manner in which transportation systems are organized in space. In a large dimension spatial model, generalized assumptions of spatial homogeneity may do little harm; as we focus on smaller scale phenomena, however, such assumptions may impair the usefulness of conclusions drawn from the model.

Our consideration of space so far has been confined to the determination of the distance *along a transportation net* between a set of points i (household locations) and a set of points j (employment sites). This spatial relationship has been labeled s_{i-j} and has had the effect of telescoping a geometrical space into a more directly relevant functional space of a single dimension. So s_{i-j} is a technological variable, for its value depends on the technology of transportation which makes movement possible. If the objectives of the model dealt only with transportation relationships between points in the system, the functional space represented by s_{i-j} would be quite adequate. However, our interest is directed to less abstract spatial considerations: because we are concerned with the location of activities and the allocation of space, the spatial framework of the model needs to be more directly relevant to the geography of urban activities.

There is need, then, to provide for the mapping of the s_{i-j} relationships at least onto a geometrical space and ultimately onto a geographical space; first, because there are spatial relationships other than purely transportation relationships that may be of interest, and secondly, because we must be able to quantify some statement about the supply of space. If we expand the distance factor by letting $_a s_{i-j}$ represent the distance between i and j via transportation system a, and $_o s_{i-j}$ represent the distance between i and j through "Euclidean" space, then we can define this mapping as

(20) $$_a s_{i-j} = T(_o s_{i.j})$$

where $$T = \text{``a transportation technology''}$$

Similarly, where the supply of land or space is involved, the Euclidean relationships between distance and area S are comparatively simple:

(21) $$S = S(_o s_{i-j})$$

so that if the translation of a relevant geography into Euclidean space is practical, the supply function can be manipulated with ease. In short, the spatial characteristics of the model are summed up in the determination of the T- and S-functions.

The T-function relates to the translation of the locational characteristics applicable to homogeneous space into a framework in which space is differentiated by characteristics of a transportation system, and vice versa. Actually, most assumptions about the "homogeneity of space" in the literature of locational economics do not describe such a space. Instead, they relate a set of spatially distributed points to a central point and say nothing about the spatial relationships between the non-central points in the set. In this sense, they describe a radial "lattice" which meets the minimum spatial requirements of the "concentric city." Lösch, however, discusses in a footnote[8] a simple T-translation between Euclidean space and a rectilinear grid: given a point origin and letting y = length of any north-south movement, and x of any east-west movement, then the locus of points equidistant from the origin *via the grid* is the square $x + y = k$, which can be inscribed in a circle of radius k in Euclidean space. More generally, the translation can be expressed as

$$_a s_{i\text{-}j} = {_o s_{i\text{-}j}}(\sin \theta + \cos \theta)$$

where θ = the angular deviation of a straight line connecting i and j from the rectilinear grid.

This is a straight space translation; another kind involves the direct or indirect functional association of space with some other variable, such as the travel time by automobile from a given point. If the relationship between time and distance were linear, the translation would be quite simple. If nonlinear, it might be very complex. Such a relationship would develop a set of "isochronic contours," such as recently described by John Hamburg:

"Consider the simple example of a hypothetical city with a limited number of radial transportation routes to the center with equal speed, Sr, on all sections of the routes. Consider also that non-radial movement travels at a speed So. Accessibility in terms of time to the CBD could be computed as the sum of the time required to reach the radial plus the time traveled on the radial, if used.

$$\text{Time} = \frac{Do}{So} + \frac{Dr}{Sr}$$

where D = distance, and S = speed.

[8] Lösch, *The Economics of Location*, p. 442, fn. 22.

"Let us assume that the settlement pattern is a function of the time distance from the city center. As the speed of travel on non-radial routes approaches pedestrian speeds or slower, a stellate pattern would emerge. If the two speeds were equal, all other things assumed equal, a concentric ring growth shape would result."[9]

To characterize the "supply of space" the relationship between a quantity of space and the price at which it may be utilized must be made explicit. Before this can be done, however, it is necessary to describe how the amount of space varies with Euclidean distances, since the T-function will permit translation of transportation cost and position rent gradients into the Euclidean framework.

In the geometry of homogeneous space the rate of change of the area S in terms of the distance s from a given point is expressed by the first derivative of the expression defining the total area:

$$dS/ds = 2\pi s,$$
where
$$S = \pi s^2.$$

In general, the supply conditions for the model are defined by dS/ds, which will be referred to as the "gross space coefficient" and represented by σ. Identification of σ becomes an obviously important part of the construction of the model, for ultimately the location of the margin point for any population will depend on the distribution of the supply of space to satisfy the demands of the population. The value of σ will observe the Euclidean expression as an upper limit, and hence

$$(22) \qquad\qquad 0 \leqslant \sigma \leqslant 2\pi s.$$

The difference $(2\pi s - \sigma)$ can arise from a number of factors, the most obvious of which are the gross geographical limitations in the supply of usable space. A city like Chicago or Cleveland with its central business district on a lake front has only half the amount of land area available in any given concentric ring about the central business district that an inland city like Denver or Indianapolis would have, and thus has a gross space coefficient more closely approximating πs than $2\pi s$. For a city such as San Francisco or St. Petersburg on a peninsula of width g,

---

[9] John R. Hamburg, "Land Use Projections for Predicting Future Traffic," *Trip Characteristics and Traffic Assignment,* Highway Research Board Bulletin 224 (Washington, 1959), pp. 76-77.

where $s \leqslant \dfrac{g}{2}, \quad \sigma = 2\pi g$

where $s > \dfrac{g}{2}, \quad \underset{s \to \infty}{\sigma} = g$

so that $g < \sigma < 2\pi g$ for all values of s.

In summary, the gross space coefficient σ is the first derivative of the geometrical expression which will quantify the *total land area* within a given distance s of the center.

Basic institutional factors associated with the city also affect the supply of land available for dwelling occupance. Sites must be served by streets which consume large proportions of the developed land in urban areas.[10] In addition, other urban uses of the land take up substantial amounts of the available space. From independent sources there is a surprising degree of agreement on the proportion of total developed space devoted to residential use, a dominant feature of the spatial structure of the city.[11] Chicago is a typical example: 40% of all developed land and

[10] "In the central and satellite cities as well as the urban areas there is a relatively constant relationship between the street area and the total developed area. On an average, central cities devote 28.10% of their developed area to this use, while satellite cities use 27.67% of their developed area for this purpose. In urban areas the average is 27.61% of the total developed area. Moreover, the proportion of land used for streets in most cities falls within a relatively narrow range." Harland Bartholomew, *Land Uses in American Cities* (Cambridge: Harvard University Press, 1955), p. 131.

Harold S. Buttenheim suggested that the average space requirement for streets in American cities was approximately 30% in the United States. National Resources Committee, *Urban Planning and Land Policies* (Washington: GPO, 1939), p. 261, Table 4. See also data in Table 2, p. 351.

Data from the Chicago Area Transportation Study suggest that the ratio of area in streets to developed area by successive distances from the center varies closely around 31% for distances between 2 and 18 miles out. Apparently there is little variation with distance except that beyond 18 miles the proportion rises somewhat, probably reflecting the scattered nature of the development. See Hamburg, *op. cit.*, Figure 1.

[11] Harland Bartholomew in his study of urban land use points out that in 53 central cities studied by him, 39.61% of all developed land was in residential use, while for 33 satellite cities the figure was 41.98%. For both groups of cities the ratio of land area in residential use to land area in all private uses was 80.2%. Bartholomew, *op. cit.*, pp. 120, 128. In six middle sized cities studied by the National Resources Committee, residential land use averaged 34.9% of all developed land use and 76% of all private land use. National Resources Committee, *Urban Planning and Land Policies*, p. 251, Table 2, p. 261, Table 4. In Detroit a recent traffic study found that 45.6% of all developed land in the study area was in residential use, or 79% of all developed land. Detroit Metropolitan Area Traffic Study, Part I, p. 38, Table 11.

80% of all such land in private use is devoted to dwelling uses; even more suggestively, beyond seven miles from the center there is very little fluctuation about the 41% level (Figure 21).

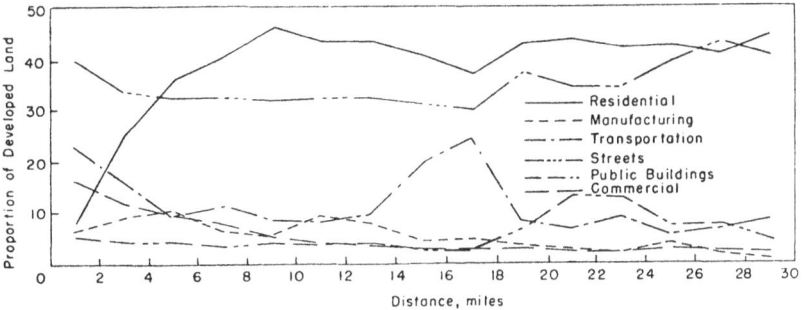

FIGURE 21. *Land use proportions, by distance from the central business district, 1956.*
Source: John R. Hamburg, "Land Use Projections for Predicting Future Traffic," *Trip Characteristics and Traffic Assignment,* Highway Research Board Bulletin 224 (Washington: National Academy of Sciences—National Research Council, 1959), p. 74, Fig. 1.

Thus a *net* space coefficient may be formulated, for where the proportion of total land area used by or available for dwellings is some function of distance s, this function applied to the gross space coefficient will yield a net space coefficient σ':

$$(23) \qquad \sigma' = \sigma'(s).$$

In a more particular sense, however, we are not so much concerned with the relationship of area to the distance from the center. The basic spatial construct is the "isotim," or equal cost contour, a locus of points about the center such that transport costs to the center for any group are equal for all such points. The relevant supply expression really involves the association of area with isotims which are not circular.

In Figure 22 the circle represents all points equidistant from the center at O. $\overline{ABCD}$ is an isotim from O via a transportation net in which $\overline{AC}$ and $\overline{BD}$ are major arteries in a grid system with uniform velocity throughout the entire system, and $\overline{AEB} - - - H$ is an isotim similar to $\overline{ABCD}$ except for the fact that nonarterial velocity is half that of the arterials. The area enclosed by each of these figures in terms of distance

measured along an arterial is, respectively, πs^2, 2 s^2, 1.33 s^2, yielding
net space coefficients of $2\pi s$, $4s$ and $2.67s$. This effect illustrates the
critical role played by the transportation system in defining the supply
of space for the model.

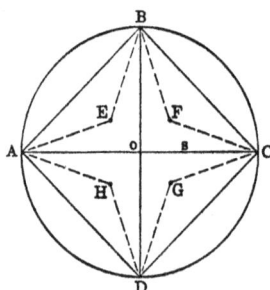

FIGURE 22.

In summary, the supply of space enters the model through the definition of the net space coefficient, which is influenced by three factors:
the gross physical limitations of the urban geography; the institutional-
technical limitations associated with the array of land-occupying urban
service and activities; and the spatial technology of the transportation
system itself in the structuring of the isocost contours in space. For a
given population a low space coefficient suggests relatively high densities and unit site rents compared to a high coefficient, for it represents
land in short supply for reasons physical, institutional, or technical.
Basic to the application of such a model is the empirical construction
of this supply coefficient.

In the succeeding chapters it will be assumed that these spatial trans-
lations are carried out for all relevant spatial conclusions, so that when-
ever s_{i-j} is referred to it will include both S- and T-translations. With
this in mind, we will feel free to use very simplified spatial abstractions
to demonstrate characteristics of the model.

There remains now one other conceptual problem. In the ultimate
we want to relate a population of households N to the structure of position rent R_{i-j}. As a prior condition, we must identify the relationship
between the population of households and the margin point m by relating
position rent R_i to the amount of land consumed per household q and
the price (unit rent) r at which it is consumed. This is the burden of the
following chapter.

V

The Demand for Space and the Basic Model

The Space Demand Function

If the demand for space by households were perfectly inelastic with respect to price, and if all households required approximately the same amount of space, the model would be quite simple to complete. Given the population of households, it would be possible to determine the total land occupied, establish the settlement margin, and define the structure of position rents. The density of settlement q_i^{-1} would be everywhere equal, and the location rent of the land at point i

$$(24) \qquad r_i = R_i q_i^{-1}.$$

This is not the case, however; the demand for space by households appears to be elastic over most of its range, so that the determination of position rents will not actually tell anything about the distribution of densities, and more, position rent is indeterminate because we can identify neither the margin point m nor the central position rent R_o upon which it depends. Thus, the model needs some component which will relate the quantity of space occupied by the household to the structure of position rents. This in turn will yield a relationship between densities and position rents which, for a given population, will establish the margin points; the density structure of the urban area will be simultaneously determined.

To review the logic of this determination of transportation costs to the household: the manifest wage rate, as has been pointed out, consists of a pure wage payment as well as of a payment to cover the location costs of the worker. These location costs include both the cost of the journey-to-work and position rent:

$$(25) \qquad W = W' + X + R.$$

Hence, if the pure wage element is reasonably constant, or at least exogenous to the model, handling the location cost premium can be viewed as a separate budget process for the worker. Thus, his budgeting problem now involves the allocation of an income L among two goods: location, measured in terms of the distance separating his dwelling and his employment site, and space, or density, conditions. The "price" of location is position rent, while the "price" of space is the unit location rent at any given point in the urban space. This problem would have a simple solution if the prices were independent of each other and everywhere the same, for the only step involved would be to map indifference curves between "nearness" and "space" and let the income line identify the optimum combination of the two that could be bought by L. However, the prices are not uniform, except circumferentially, and the two transactions are related by the simple definition that position rent is the product of the quantity of space consumed and the price (or location rent) at which it is consumed:

$$(26) \qquad\qquad R = rq.$$

Here r and q are indeterminate, and so it is necessary to turn our attention to the individual demand for space.[1]

Empirically we know little about the manner in which consumers' preference for space (or conversely, for density) is related to the price of space. The mechanism of the housing market is so complex that the nature of the household's demand for space tends to be obscured in the intricate bundle of products called "housing." In buying "housing" one is purchasing a number of things: living space, physical and service amenities, location, prestige values, and so on. In the market for the bundle, that part of the price which is actually being paid for space and location is difficult to isolate. In a gross sense these are factors weighed by consumers at the time of the transaction, and so there is some justification for proceeding with the intuitively best assumptions

[1] Using three products, location, land, and a composite of all other consumption possibilities, Alonso defines a "bid price curve" for land: "a set of combinations of land prices and distances [from the center] among which the individual is indifferent" and uses it as a parametric function because it permits "a solution which combines the indifference curve approach with an explicit consideration of land prices." For any individual, equilibrium exists when the individual locates at the point at which "the price structure touches the lowest of the bid price lines with which it comes in contact. . . . [Here] the price structure curve and the bid price curve must be tangent. . . ." Alonso does not rely on a formal demand expression, for this is subsumed in the "bid price curve." William Alonso, *A Model of the Urban Land Market* (unpublished Ph.D. dissertation, University of Pennsylvania, 1960), pp. 127-29.

that we can make concerning the nature of the relationship. To begin with, assume that the price of all other goods is given and constant, that the household demand for space is uniform throughout the population, and that space has a diminishing marginal utility for the household. The space demand function will have a negative slope: the greater the unit rent the fewer units of space will be consumed.

On Figure 23 is plotted a space demand curve *aa* and position rent lines R_1, R_2, etc. Nothing is said about distance from the center; R_1 has reference only to the distance from the margin point to s_1, as in Figure 24. Now follow the relationship between the margin point and the central point. Every household will operate at some point on the space demand curve *aa*. Note that although only a few R_i lines are shown, the entire field is structured by continuous values for R_i—every point that lies on *aa* lies also on some R_i curve. Recall also that R_i is a variable dependent on the distance from the margin point *m* to point *i*, as indicated in Figure 24.

The effects of population growth can be easily simulated with reference to these two figures. Letting *o* represent the central employment site and *m* the current margin of settlement, and assuming that the transportation function X_{o-m} is constant for all users, then we can simulate the growth effect by extending the margin in Figure 24 from *m*

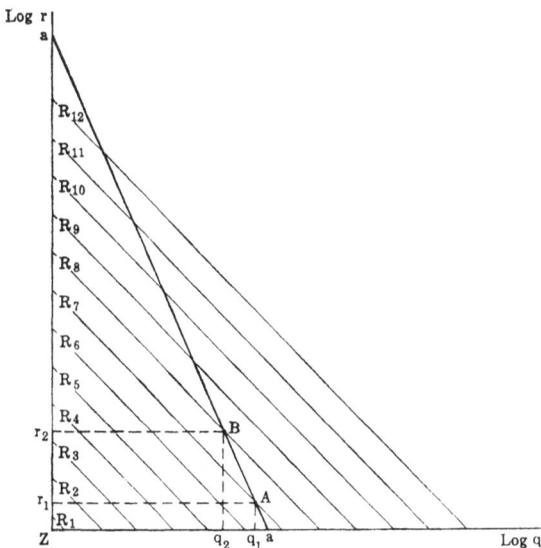

FIGURE 23.

to m'. This results in an upward shift in position rents from R_7 to R_8 for any household domiciled at i, so that in a dynamic sense the result of urban growth is a continuing increase in location costs throughout the region, unless transportation improvements take place at a rate which would permit the decline in the position rent gradient to offset them.

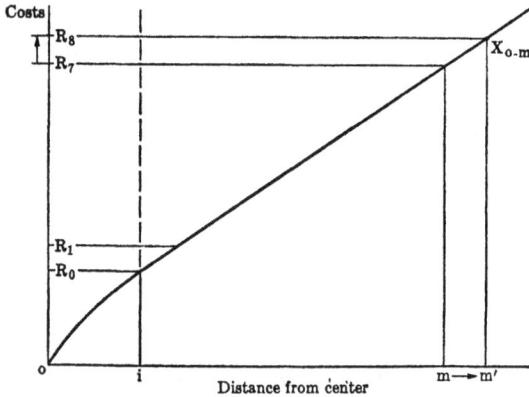

FIGURE 24.

The long run growth effects are indicated in Figure 23. Assuming that the demand schedule is constant, the individual at i would counter the rising location costs by shifting along his demand curve from A to B, that is, he would reduce his consumption of space from q_1 to q_2, while the location rents per unit area rise from r_1 to r_2. In other words, the long run consequences of urban growth will express themselves first in the form of rising land values and second in the compensating consumer effects of increasing density generally over the entire urban region; that is, as q declines, density, or q^{-1}, will increase.[2]

This simple example serves to illustrate the interaction of the margin point m, which establishes the position rent structure for the urban area. Position rent, in turn, identifies the point on the demand schedule of the individual at which prices and densities are in equilibrium. Here he is restrained by the total pattern of his consumption preferences from

---

[2] This is essentially "net residential density," the relationship of a population to the total space used for residential purposes. It is, however, adaptable to any other definition of population density. Cf. William H. Ludlow, "Urban Densities and their Costs: An Exploration into the Economics of Population Densities and Urban Patterns," in Coleman Woodbury (ed.), *Urban Redevelopment: Problems and Practices* (Chicago: University of Chicago Press, 1953), pp. 107-112.

paying any more than r_i, and the competition of other users will prevent his paying any less, once the value of the position rent is established.

For the purposes of the model, then, it will be assumed that the space demand function is constant among all households and is logarithmically linear in shape:

$$(27) \qquad q_i = \left(\frac{\lambda}{r_i}\right)^{\frac{1}{\eta}}, \quad \eta > 1$$

where q_i = the quantity of land consumed per household at point i,
$\quad r_i$ = the unit value of space at i, and
$\quad \lambda, \eta$ = parameters of the demand function.

This expression becomes more directly useful for the model when q_i is expressed in the form of density q_i^{-1} thus:

$$(28) \qquad q_i^{-1} = \left(\frac{r_i}{\lambda}\right)^{\frac{1}{\eta}}$$

Moving away from the most simple form of the model, we can elaborate the role of demand in two major directions. In the first place, demand functions will be unique to households, so that we can write the parameters in more specific fashion—$_k\lambda$ and $_k\eta$—to express the demand function for household k. In the second place, variations in time preferences will lead to the condition that for any point i each household will experience a unique position rent and hence will be willing to pay a unique unit rent, so that these rents can be expressed more specifically as $_kR_i$ and $_kr_i$. To identify these specifications we will define the *specific form* of the demand conditions as follows:

$$(29) \qquad {}_kq_i^{-1} = {}_k\left(\frac{r}{\lambda}\right)^{\frac{1}{\eta}}{}_i \,,$$

although for the exposition of the model in its most simple form we will refer only to the general form above.

Assembling the Basic Model

Definition of the individual space demand function provides the last element necessary for the assembly of the model in its simplest form, for the mathematical expressions which have been described offer a linked set of relationships which can describe a set of location rents and densities of occupance for the urban region, given a population of certain defined characteristics, a described spatial framework, and a small set of miscellaneous variables.

As a first step the demand function can be manipulated so that it will express density q^{-1} as a function of position rent R:

$$q = \left(\frac{r}{\lambda}\right)^{\frac{1}{\eta}} \qquad \text{The demand function, } \eta > 0$$

$$r = Rq^{-1} \qquad \text{Position rent in the sub-budget.}$$

Then

$$(30) \qquad q^{-1} = \left(\frac{R}{\lambda}\right)^{\frac{1}{\eta-1}} .$$

Since the demand curve is negatively sloped and since the position rent curve is negatively sloped with respect to distance, it follows that the density gradient must be negatively inclined with respect to distance from the center. Relating the density gradient to a given population, then, is analogous to finding the limits of integration necessary for a negatively sloped function rotated about an axis to describe a specified volume in Euclidean space:

$$N = 2\pi \int_{0}^{m} sq^{-1} \, ds .$$

For purposes of this model, the term $2\pi s$ is replaced by the net space coefficient σ' described earlier, so that the primary function of the model can be expressed generally as

$$(31) \qquad N = \int_{0}^{m} \sigma' q^{-1} \, ds .$$

The purpose of this expression is to determine the value of s at m, which, once defined, will provide for every value of $s \leqslant m$, that is, for all i, a value for density q_i^{-1} and for location rent per unit of space r_i.

The position rent element R can be extended in the following fashion:

where $R_i = X_{m-o} - X_{i-o}$ Definition

and $X_{m-o} = \phi(s_{m-o}, N)$ Transportation cost function between the margin m and the center o depends upon the distance between them and the transportation demand features arising from the population N. Other transport inputs and prices assumed constant.

and $X_{i-o} = \phi(s_{i-o}, N)$ See above

then $R_i = \phi(s_{i-o}, N)$ (S_{m-o} is a special case of S_{i-o})

and the full summary expression for the basic model can be set out as

$$(32) \qquad N = \int_o^m \sigma' \left[\frac{\phi(s_{i-o}, n)}{\lambda} \right]^{\frac{1}{\eta-1}} ds \quad .$$

This is the critical expression for the model, for it identifies the margin point m upon which the absolute level of position rent depends. As the dependent variable, m is related first to population N, this relation has a number of consequences for the model. Its most direct effect is manifest in its characterizing the number of consuming units in the market for urban space available for residential use. In another guise it sets the level of deadline demand upon the transportation system and thus influences the time-costs of spatial separation. If we expand the system to envelop the labor market, N is related to employment characteristics of the labor market and hence itself depends on the demand curve for labor in the local economy. When we depart from the simple assumptions about the homogeneity of the population, the effects of N enter the model by way of the mix of critical characteristics of the members of the population as consumers of space and of the services of transportation.

In the second place, m depends on the supply characteristics of space about the central point, which are summarized by the net space coefficient σ'. In terms of this expression, σ' defines the extent of the rotation of the density gradient for every value of the distance s from the center. As previously defined, σ' has two distinct components: on the one hand it embraces the purely physical limitations on the supply of space, which are occasioned by the natural geography of the urban hinterland; on the other hand, it brings into the model the institutional limitations on the availability of space, including existing occupancy of the space, policy restrictions, and other quantifiable social features of the land. It should be kept in mind that σ' also has a technological dimension, for its distance element s is actually the transformation of Euclidean distance into effective distance in the transportation system. Thus, σ' really expresses the increment to the supply of space made available by increasing the distance from the center by a unit of distance which is very small. Basically, σ is an empirical function to be measured in the real world and then described by a fitted mathematical expression.

The demand characteristics of the land market enter the expression through the two parameters η and λ which describe the slope and position of the individual space demand curve. These are taken as uniform for the entire population here, but where the population is disaggregated by consumer characteristics the entire margin expression would have to be modified to relate to the specific form of the demand function discussed above. This suggests a proliferation of computational complexity, for the market dominance of the separate groups over the range of s would have to be identified by determination of the points at which the unit prices offered by the group demand curves are equal. What is involved here is closely related to agricultural location theory describing the spatial characteristics of production where a single market point and several crops are the principal elements.[8]

The core of the margin point expression is the transportation cost element $\phi(s_{i\text{-}o}, n)$. This element embraces the major technological characteristics of the urban transportation system and thus determines the spatial framework of the model. These technological characteristics are translated into time-costs, and these in turn are transformed into value costs through the pricing mechanism of the marginal value of leisure, so that the ϕ-function is in large part the subjective consumer appraisal of the critical objective characteristics of the system.

Thus, the margin point expression emerges from the integration of all of the major elements into a single model expression. With the determination of m, the position rent surface R_i is everywhere determined by

$$R_i = X_{m-o} - X_{i-o}$$

---

[8] Edgar S. Dunn, Jr., *The Location of Agricultural Production* (Gainesville: University of Florida Press, 1954), and J. H. von Thünen, *Der Isolierte Staat in Beziehung auf Landwirtschaft und Nationalökonomie* (Hamburg, 1826).

and for every value of R density is determined

$$q^{-1} = \left(\frac{R}{\lambda}\right)^{\frac{1}{\eta-1}}$$

as well as the unit rent

$$r = Rq^{-1}.$$

Thus we have arrived at a statement of the model in its simplest form. It is simple insofar as it deals with homogeneous aggregates: the spatial framework is uniform in all respects except in terms of distance from the center. The population of households is uniform in the important respects of the marginal value of leisure and the space demand function. It has optimizing features in the sense that it describes a partial equilibrium structure which, given the valuation put on leisure and the space preferences of the population, represents *ceteris paribus* the distribution of rents and households which maximizes the welfare of each.

The usefulness of this simple form of the model is its ability—given the relevant descriptions of (1) the technology of transportation, (2) the economic valuation of leisure time as it relates to the journey-to-work, (3) the valuation of residential space by consumers, and (4) the supply characteristics of urban space—to generate a structure of location rents (and thus a structure of land values) and densities in terms of the major aggregate—population. Taken in another way, it relates the movement of the margin point to changes in population and hence is suggestive about the rate at which land tends to be converted to urban use. As a partial equilibrium model it has another significance. It suggests welfare, or efficiency, criteria for policy relating to the spatial structure of the urban household economy: given the functions and assumptions expressed, departure from the conditions expressed by the output of the model represents loss in welfare for the community as a whole. In the following chapter, some of these implications will be explored, along with the effects of some alternative assumptions.

VI

Applications of the Model

The Model and the Real World

Before we examine some of the implications and applications of this model of urban structure, a summary review of the major theoretical assumptions and elements may be fruitful in setting up a perspective about the relationship of the model to reality. The basic elements of a locational model can be identified as

a space which is economically differentiated in some definable fashion, usually by costs of transportation necessary to overcome spatial separation of activities, and
a set of activities for which this special spatial differentiation has economic consequences.

Given these features and the behavioral axioms that

the "owners" of the space will seek to maximize returns from the employment of space, and
all firms (households) will seek to maximize net returns,

there is an equilibrium distribution of these activities in space such that

no firm (household) can increase its net returns by relocating, and
no owner of space can increase his returns (rent) by getting other than the present firms or activities to occupy his space.

This condition has been referred to as "locational equilibrium": not only is there no economic motivation for any unit to change the pattern, but the net social returns are maximized.

This study has applied the locational framework to the spatial structure of the city, concentrating on a sector of special importance in the sheer volume of its space requirements—the household sector. It has treated two sorts of activities: those producing goods or services which set up the demand conditions for the employment of the labor force, and those "producing" labor services as inputs to the first carried on by a set of "firms" (households).

Space is differentiated by the costs of transporting the labor input to the sites of production; more specifically, by the spatial distribution of production sites and the technology and organization of the transportation system. These, in turn, determine the costs of movement from any point in the region to the employment sites.

We assume that no worker will be employed at a wage less than his opportunity costs as measured by (1) competing employment opportunities and (2) the value which he places on his leisure. This assures that he will always accept employment at the employment site offering the largest *net* return to him provided that the daily wage paid is at least equal to his valuation of the leisure given up. Given the manner in which the worker values his leisure, the time-costs of transportation are expressed in many terms; with work trip outlays they establish a structure of position rents. Position rent is simply the economic advantage in transportation costs of any location with respect to the most disadvantaged position occupied, that is, the maximum amount that a user would be willing to pay rather than occupy the marginal location.

Assuming that households exert a demand for space for residential purposes which reflects the principle of diminishing marginal utility, then, given the structure of position rents, each household will choose a location which will maximize its net satisfactions. As it moves toward the employment site, position rent increases and transportation costs decrease until it finds a position in which the marginal saving in transportation costs is just equal to the marginal value of the residential space given up.[1] Through this process of substituting transportation costs for space costs, locational equilibrium is achieved. In short, given

(a) the spatial pattern of employment centers,
(b) the organization and technology of transportation,
(c) a population of households,
(d) the marginal valuation of leisure by the worker, and
(e) the marginal valuation of residential space by the households,

the spatial distributions of densities and rents under conditions of locational equilibrium are simultaneously determined, as is the amount of land required for residential use, its distribution, and its value.

---

[1] Or to the marginal disutility of the enhanced density conditions.

The very simplicity of this summary suggests the degree of divergence that exists between the model and the "real world." In the first place, the conditions of perfect competition are most certainly not realized in the real world markets of interest to us. The legal qualities of the land, its marketability, the structure of its ownership, the rigor of public regulation, the organization of the housing market and construction industry are sources of important institutional frictions involved in adjustments in the direction of an optimal spatial pattern.

In the second place, the locational decision of households is not nearly so simple as the model suggests. In addition to the journey-to-work, the choice is influenced by the distribution of the stock and quality of housing, prestige and culture group associations, variations in the quality of highly valued local services, such as schools, and many other considerations. Complex substitution effects are certain to take place among these if rational behavior asserts itself at all. Some of the advantages of nearness to employment may be given up for a neighborhood of greater prestige, or a household may pay a higher rent in order to enjoy access to an unusually good school. The importance of such considerations in the locational decisions of individual households is not to be depreciated, but this study has chosen to focus rather on the manner in which some critical technical variables influence the spatial organization of the urban community through their effect on the individual decision in the market; it defines an optimum arrangement for all where these other features are either generally distributed or of little influence in individual decisions.

Finally, in relating the important variables to space we have expressed them in terms of the distance from some focal point, so that we seem to have been preoccupied unduly with a linear, or at best a radial, construct of population distribution and the land values associated with it. This has left us in the position of saying nothing about the circumferential distribution of the variables, that is, about how the variables are distributed along (as distinguished from "among") the iso-lines of the space. Since transportation costs of the journey-to-work will not distinguish among such points, one can speculate that it is in this aspect of distribution that the special factors are decisive. In this sense, there is no real incompatibility between concentric and sectoral constructs. For example, better quality land in one *sector* of the circle may attract higher income groups who are willing to bid the land away from lower income groups, so that a high income, prestige area gets its beginning and attracts other high income households. The advantageous physical endowment of the land may lead ultimately to a "social endowment" which in itself is highly valued by certain social groups in the community; the regular concentric structure of distribution is distorted and a sectoral structure emerges. This is precisely the nature of the phenomenon described and analyzed by Homer Hoyt which has become known as the "sector" hypothesis of urban organization.[2] Thus, these specialized factors may result in substantial modifications of the concentric form without abridging the effect of the more general variables in setting the basic conditions of urban settlement for the great majority of people.

In spite of these divergences between the model and the real world, to understand something about the abstract characteristics of the urban structure which is theoretically optimal in the distribution of population over the urban landscape can be quite important in suggesting directions for and limitations of public policy regulating the spatial organization of the city. Specifically, public policy seeking to modify the urban structure by manipulating such technical variables as the transportation system and the supply conditions of the market for land will succeed only by accident unless it considers the time and space preferences of the population. Thus, the impact of land use policies seeking to mold growth into a more compact urban form is frequently dissipated by the powerful effects of transportation improvements which reduce position rent gradients and expand critical isochrons[3] far into the hinterland, encouraging dispersion rather than concentration. The end result of such a conflict must be something less than an optimum arrangement of the urban structure. An important value of a model such as this, therefore, is that it affords a framework in which certain critical behavioral elements can be seen interacting with the salient technological dimensions and policy features of the urban environment. So long as the market, circumscribed as it may be by public policy, is the principal machinery allocating urban space among competive uses, this interaction will be a dominant city shaping force in our society.

With the completed model some variations on the theme can be explored. We will avoid any rigorous formulation and describe only

[2] Homer Hoyt, *The Structure and Growth of Residential Neighborhoods in American Cities* (Washington: GPO, 1939).

[3] The position rent gradient is the amount of saving in transportation costs at any point *i* over the transportation costs at the margin point *m*. An isochron, again, is a contour in space of points of equal transportation cost with respect to the center.

the general nature of the effects to avoid greater strictness than our knowledge of the variables justifies. We can, of course, measure and quantify the technical variables with considerable precision. We cannot describe the specific content of the economic functions at this stage in our knowledge and must rely on economic intuition to suggest their more general characteristics. For purposes of demonstration we shall frequently condense two dimensional space to a linear form by expressing the variables as some function of distance, $f(s)$. Here $f(s)$ really stands in the place of some two dimensional expression, such as the radial form $2\pi f(s)$.

We can investigate the applications of the model in three directions: (1) the market effects: analyzing the consequences of changes in the parameters relating to the market behavior of the population; (2) the policy effects: analyzing the consequences of certain kinds of centralized decisions by public agencies which affect the technological conditions of the model; and (3) the welfare implications: analyzing how to make better decisions through the use of such models.

In the first case, some of the gross assumptions of the basic model will be modified in the direction of more realistic characteristics. In the second case, certain kinds of public decisions will be simulated to suggest some policy uses of this and similar models. Finally, with a cue from welfare economics, the model will be applied to some decision situations in which the focus will be on changes in social costs and benefits. Bear in mind that even at its most elaborate, this model is circumscribed with simplifying assumptions; more elaborate refinements would amplify the complexity of effects, perhaps, without disturbing these more gross conclusions.

The Effects of Decentralized Decisions—The Market

To recapitulate, the role of the economic functions in this model can be described in the following propositions:

Given the technical conditions, the individual's marginal-value-of-leisure curve will establish for him a distribution of position rents $_kR_{i-j}$ with respect to any margin point m, and

given the distribution of position rents $_kR_{i-j}$, the individual's space demand curve will describe his indifference surface with respect to site rents, density, and distance from the employment focus.

To express this relationship another way, since location costs L are constant for the individual for all points under the posited assumptions, equilibrium for the individual results from the substitution of transportation costs for densities up to the point where the value of the marginal gain in space (density) conditions is equal to the marginal distance cost of transportation.

Although we have assumed up to this point a population of households homogeneous in terms of how space and leisure are valued, we have every reason to believe that there is considerable variation, in any urban population, in the subjective valuation of both residentiary space and the time consumed by the journey-to-work. We might divide all members of a population into two classes with respect to each of these variables: high and low. This yields four groups (see Figure 25).

FIGURE 25.

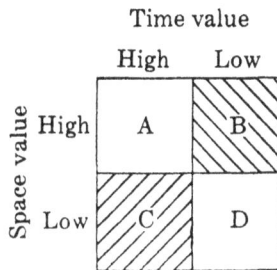

Intuitively we conclude that the extreme cases are B and C. B is a centrifugal class in that its high individual demand for space is reinforced by a low valuation of the time-costs of work-residence separation; these circumstances tend to bring about a low density, dispersed type of settlement spread out thinly across the countryside surrounding the employment focus. In this respect the B class is analogous to our "exurbanites" —the upper middle income families with school age children.

The C class is the "centripetal" class: the low individual demand for space, augmented by a high valuation of the time-costs of the journey-to-work, leads to a relatively high density clustering about the central employment focus. Using the same analogy we used for the B class, we might compare the C class to "urbanite," the upper middle income childless family who chooses to live in high density apartments adjacent to the downtown area.

The A and D classes are conceptually ambiguous—their locational characteristics will be governed by the result of the centrifugal and centripetal features of the economic variables working against each other. High time-cost valuation will cause the A class to tend toward concentration at the center, but the higher individual demands for space will tend to counteract that tendency. In the D class it is low space demand that causes a tendency toward concentration, while low time-cost valuation is the counteracting force for dispersion.

The ultimate equilibrium pattern among all these groups would depend not only on these explicit economic functions but on the population and income levels, which set the aggregate demand conditions. Thus any investigation of the effects of variations must make some statement concerning them.

To simplify manipulation of the model, assume a population consisting of only the two extreme classes, B and C discussed above; incomes of all members are equal. Let the total population P be divided into these two groups so that the number of members of the B group P_B is equal to some proportion b of the total population, thus:

$$P_B = bP$$

and likewise

$$P_C = cP$$

so that

$$b + c = 1.$$

Both groups face a common set of technical conditions: all labor force members of both groups are centrally employed, use the same transportation system, and compete for the same space. We can summarize the technical conditions by describing the common transportation function as in Figure 26.

In this function FH we have assumed a constant operating velocity represented by the constant slope of FH, and an average ingression loss, all of which occurs at the center, represented by OF. An alternate assumption would distribute ingression losses through at least part of the system, yielding a function such as OGH, which implies a decline in average velocity as one moves from G into increasing congestion until the center is reached. It is the OGH type of function which will be assumed hereafter.

The C group is distinguished from the other in part by the fact that its members place a high value on time spent in transportation; in other words, the slope of the transportation cost curve for C group members

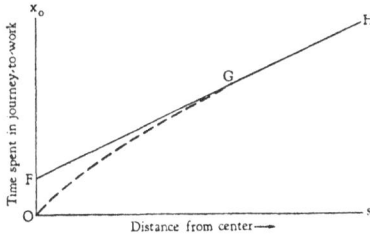

FIGURE 26.

is greater than that for *B* group members. Thus when the technical transportation function in Figure 26 is translated into cost terms for each group, each group will act locationally under different cost conditions as indicated in Figure 27.

It further follows, given these cost functions, that for every value of the central rent R_o each group has a separate and distinct position rent curve, such that the position rent curve of the *C* group has a steeper slope than that of the *B* group.

Now, if each group has a different space demand curve, such that each member of group *C* will pay more for space than each member of group *B*, then for a given value of position rent R_i members of the *C* group will demand less space and be willing to pay more for it than members of the *B* group.

At this point we can express these general considerations in graphic form (Figure 28).

The upper part of Figure 28 represents the distribution of unit rents

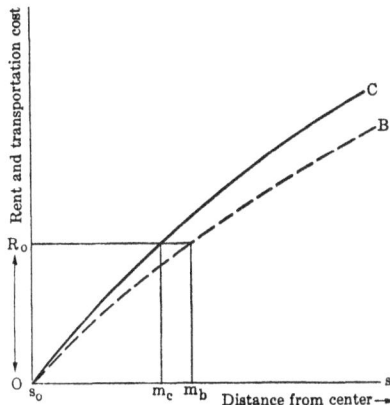

FIGURE 27.

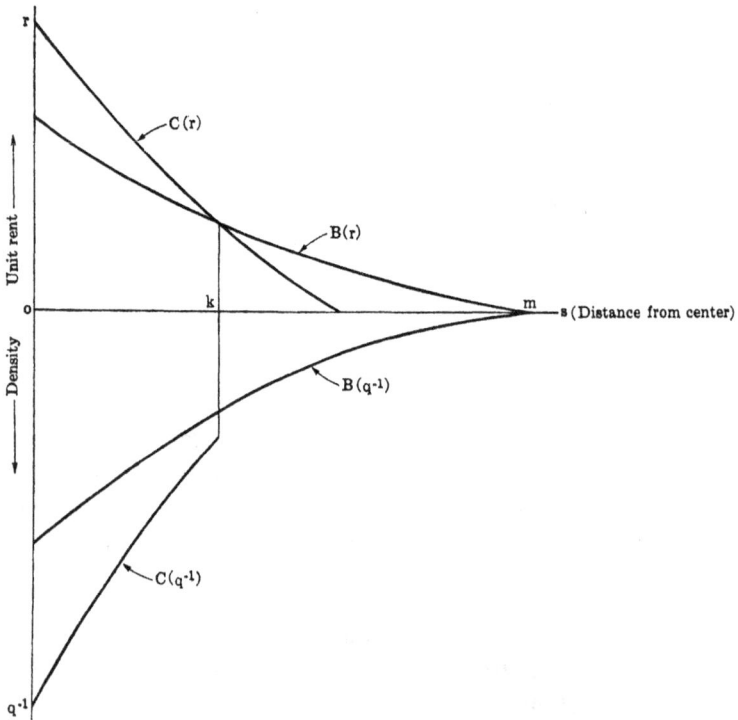

FIGURE 28.

from the center point o to the margin at m. The lower part indicates the distribution of densities in the same space. k is the inner margin marking the boundary between group C, distributed according to its density function $C(q^{-1})$ along its unit rent curve $C(r)$, and group B, distributed according to its density function $B(q^{-1})$ along its unit rent curve $B(r)$. This represents an equilibrium distribution *between the two groups* under the following conditions:

At point k

$$(1) \qquad\qquad B(r) = C(r)$$

and (2) $\qquad \int_o^k \sigma[C(q^{-1})] = cP \text{ and } \int_l^m \sigma[B(q^{-1})] = bP.$

It represents an equilibrium among the members of each group if for any point i,

(3) $$C(r) = \theta[C(q^{-1})]$$
and $$B(r) = \psi[B(q^{-1})]$$

where θ and ψ represent the space (density) preference functions of C and B respectively.

And (4) $\quad R_{i,C} = C(r_i) \times \theta[C(qi^{-1})], \quad R_{i,B} = B(r_i) \times \psi[B(qi^{-1})]$

where $R_{i,C}$ and $R_{i,B}$ represent the position rent at any point i for C and B respectively,

(5) $\quad R_{i,C} = R_{o,C} - T_{i,C}$ and $R_{i,B} = R_{o,B} - T_{i,B}$

where $R_{o,C}$ and $R_{o,B}$ are the central position rents for C and B respectively, and $T_{i,C}$ and $T_{i,B}$ are the transportation costs between o and i for C and B respectively.

In summary, given the technical conditions, the class distribution of population, incomes, the manner in which the time spent in the journey-to-work is valued for each class, and the household space demand for each class, there is an equilibrium distribution of population and land values in the urban space which will maximize the net satisfactions of all members of the population. This distribution meets the requirements for location equilibrium because (a) no household can improve its positions by relocating (conditions 3 and 5 supra), and (b) the space is so allocated as to maximize its return (conditions 1 and 2 supra).

This case can be theoretically extended to any number of groups exhibiting variations in space and time preferences. Since the spatial distribution of households is associated with the pattern of journey-to-work isotims, and hence with the technical conditions obtaining in any given case, the practical result is a generally concentric pattern of distribution. In this sense, the model is consistent with Edgar Dunn's elaboration of von Thünen's model of agricultural location and Richard Muth's application of the von Thünen model to the rural-urban fringe,[4] since all deal with the patterning of land-extensive uses which results from the effect of transport costs on the net returns position of the competing units. In lieu of unit costs of transporting an agricultural product, the urban case uses a cost concept of the journey-to-work; its net prod-

[4] Edgar S. Dunn, *The Location of Agricultural Production* (Gainesville: University of Florida Press, 1954), see especially "Maximation Solution for Two Products" (pp. 9-13), which is the agricultural analogue to our case. Richard F. Muth, "Economic Change and Urban vs. Rural Land Conversions," *Econometrica,* January 1961, Vol. 29, No. 1, pp. 1-23.

uct involves wages net of transport costs and the satisfactions associated with residentiary space (density) conditions. The individual household seeks to maximize its net satisfactions by substituting transportation costs for density.

One more observation is in order. Although competition in the market for space will tend to yield a fairly regular gradient of unit rents from the center, the same thing cannot be said of densities where component groups of the population are extremely differentiated in terms of the basic valuation of time and space; in this case the density gradient may tend toward sharp discontinuities at the internal margins between groups. Where such extreme differentiation does not exist (in other words, where our two extreme groups are joined in the population by more ambiguous groups such as *A* and *D*), a more continuous density gradient can be expected.

The technical conditions of the model are essentially summarized by three features:

> the transportation technology, as subsumed in the transportation function, i.e., in the relationship of distance and time consumed in movement typical of the technology operating at peak periods,
> the "supply" of space in terms of distance from the employment center, and
> the distribution of employment points in space (hitherto assumed to be concentrated at a single point).

The "shape" of the urban structure will be considerably influenced by departures from, or policy modifications of the conditions hitherto posited. To provide some idea of the manner in which variations in these features tend to shift distribution of densities and land values, on the succeeding pages we will examine some variations in these basic technical conditions.

Variations in the Transportation Technology

There are three types of general effect that suggest themselves: those resulting from "growth," or the exogenous increase in deadline demand; those resulting from an increase in system velocity; and those resulting from general improvement of the system.[5]

---

[5] At a later point some effects resulting from certain changes localized within the system will be examined.

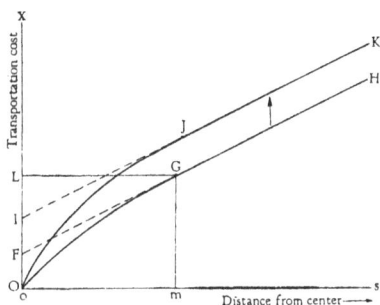

FIGURE 29.

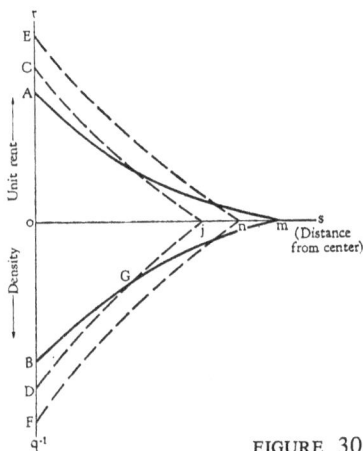

FIGURE 30.

The basic transportation function is illustrated in Figure 26. It consists of two components as discussed in Chapter 2: the system velocity (the rate at which units tend to move when unimpeded by capacity chokes), which is represented by the slope of line *FH*; and the ingression component resulting from the "loading" of the system, represented by the intercept *OF* of *FH* with the vertical axis. In a system of variable capacity, ingression tends to be distributed in the system in the form of congestion losses, so that the curve *OGH* is probably more representative of the transportation function as generally experienced in urban areas.

The growth effect may result from a number of developments. The most obvious of these is an increase in employment at the center, the immediate consequence of which is to increase deadline demand and thus the amount of ingression loss in the system. Since system velocity is not directly affected, the basic slope of the function would remain unchanged, and the ingression effect would be demonstrated by a shift upward of the function to a higher intercept on the time axis, as from *OF* to *OI* in Figure 29. Under pre-growth conditions, location costs were equal to *OL* with the external margin at *m*; the rent and density distributions are shown in Figure 30 as *Am* and *Bm* respectively. The major effect of growth on the transportation function is to steepen the function as it approaches o, with the concomitant effect of steepening the position rent curve. This leads immediately to a new set of steeper unit rent and density curves. Now if we redistribute the pre-growth population in accordance with the new unit rent and density curves, we get the new density distribution D_j and a new unit rent distribution C_j. Since the population is measured by the area under the density curve, the area B_mO must equal the area D_jO. This is the minimum condition, positing zero growth. Thus, actually the new margin point must fall to the right of j at some point such as n; the area $D_{jn}F$ would represent the amount of population growth. Several observations follow:

> Since growth increases the slope of the transportation function, it will almost certainly increase both unit rents and densities around the center.

> It is possible that, as an unanticipated effect of growth, values and densities will fall at the margin; indeed, that the margin itself will contract if the ingression effect is sufficiently pronounced and the increment to population is sufficiently small.

> Thus, this model contradicts any *ad hoc* assumptions that extension of the urban margin results *pari passu* from population growth.

> The extent of this inward redistribution could be moderated by a public investment program aimed at preventing any increase in ingression levels by adding capacity to the transportation technology.

The Distribution of Employment Foci in Space

Up to this point we have assumed a single employment center, not only for geometric simplicity but because it is generally characteristic of larger cities that the preponderance of employment does take place in or adjacent to the central business district. We would be remiss, however, to ignore cases in which employment may be distributed in a number of centers, for this situation may have a substantial distribution effect of its own. Positing conditions of full employment and locational equilibrium about the central employment point, we can observe the manner in which a new employment center will disturb that equilibrium, the sequence of the adjustments to a new equilibrium, and the nature of the new equilibrium after the adjustments have worked themselves out.

The full employment assumption is realized when the supply SS and demand DD curves for labor are equated at a point which exhausts the labor supply Q, as indicated in Figure 31. This condition exists at some wage W. Given a uniform transportation function and the wage W, a structure of position rents $R_o m$ is produced (Figure 32) such that the spatially marginal worker at m will be indifferent between accepting

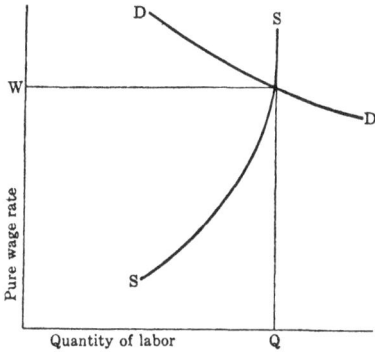

FIGURE 31.

employment at the center and remaining idle. Thus is produced a spatial distribution of density Dm and unit rents (not shown) through the assertion of the space demand of the population.

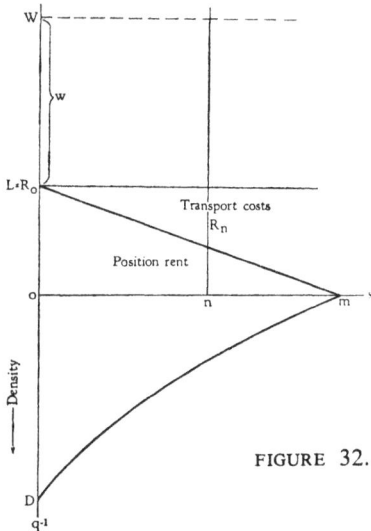

FIGURE 32.

In Figure 32 the total population N:

$$N = \int_{o}^{m} \sigma' \, q^{-1}(s) \, ds$$

Now assume that a new point of employment emerges at n. To equate pure wage rates, it follows that if manifest wage rate W offered at n is less than the pure wage rate w plus the position rent R_n at that point, no members of the labor force will accept employment at n, for the position rent structure so generated would be everywhere lower than that established by the central point and the wage W. This would mean that anyone employed at n would have to pay the higher rents established by the center so that the pure wage w would be less at n than at o, a condition incompatible with locational equilibrium.

Hence a wage would be established at n such that

$$W_n > w + R_n$$
$$\text{or}$$
$$W_n = w + R'_n, \; R'_n > R_n$$

The first consequence is that it now becomes advantageous for all workers residing to the right of j in Figure 33 to shift employment to the new focus, for they will receive a wage premium equal to the difference between the old position rent curve $R_o m$ and the new $R_j R'_n m'$. This consequence we will label the "employment shift effect."

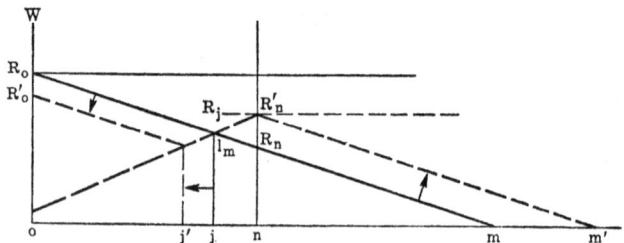

FIGURE 33.

The second consequence, resulting from the competition among households for position about point n, is conversion of this wage premium into unit site rents. Households in the o-j sector will tend to move into area j-m, switching employment to n to seek to capture some of the wage premium. This effect we will call the "domicile shift." There will then be a decline in the central position rent curve from an intercept at R_o to one at R_o', with a new intersection of the two position rent curves at j'. At this point the pure wage rates at o and n are equated at a level higher than the initial level by the amount R_oR_o'. The net effect is a reduction of densities and land values between o and j and an increase in them between j and m'. The part of the labor force residing to the left of j' will continue to find employment at o, while that to the right of j' will be employed at n as indicated in Figure 34. The area $D_o'm'r_o'$ represents an equilibrium in the distribution of densities and

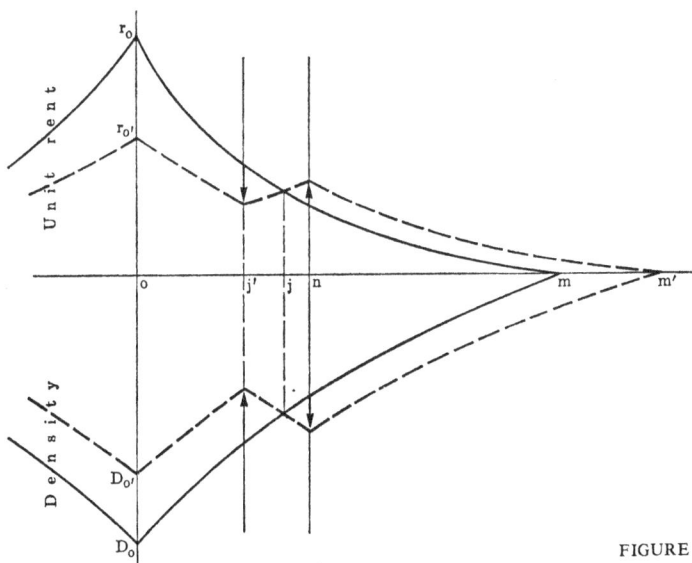

FIGURE 34.

unit rents. It is a locational equilibrium since no household can improve its position by relocating.

Equilibrium, however, does not exist in the labor market, for the capture of part of the labor market by the new employment center at n has had the effect of reducing the labor supply available to the employment focus at o to Q_1 as indicated by Figure 35. The labor force employed at n will hence be equal to Q-Q_1. It now follows that wages at o will tend to rise along the demand curve D_o-D_o and will be countered by rises in the wage rate paid at n along its demand curve until a point w' is reached where the total labor force Q is so divided among both centers (as at Q_2) that the division of the market equates the joint supply with the joint demand $D_o'D_o'$ at a new combination of wages paid at the two points. This will result in adjustments in the postion rent curves of each with concurrent adjustments in the spatial distribution of densities and unit rents. Hence the achievement of a new equilibrium in the labor market will result in a new locational equilibrium. The nature of the labor market equilibrium will be conditioned by the elas-

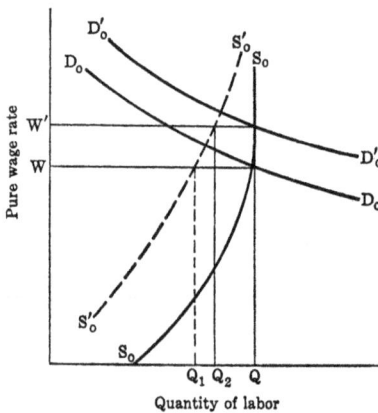

FIGURE 35.

ticity of the demand curves associated with each of the employment centers. In short, if we are given the spatial relationship of the two centers, the transportation function as it faces the labor force, and the economic functions of the marginal valuation of leisure and the space (density) preferences of the labor force, the equilibrium distribution of households and land values is simultaneously determined by the labor demand curve associated with each employment focus. This case illustrates in simple form the interaction of basic economic relationships and spatial relationships characteristic of urban organization and the extent to which the effects of a spatial change are diffused through a network of economic interaction before becoming manifest as spatial phenomena.

Policy Effects of the Model

Public policy enters the model by way of its effect on supply conditions in the market for residential space. On the one hand, public agencies may take certain kinds of action which influence the effective space coefficient σ'; these actions are brought about by the authority of the agencies to regulate the manner in which urban space may be used, and by public policy which creates and allocates "social overhead"— those public improvements which amplify the usefulness and productivity of land used for siting urban activities. Although land use regulation may enlarge the supply of residential space by restricting its use by competing activities, probably its more general effect is to limit the supply. It may do this directly by barring space from use by residences or by establishing minimum site requirements which limit the intensity with which space may be occupied. Such restrictions generally work toward lower gross densities or higher aggregate land requirements, pushing the margin farther out into space and raising location costs.

Public overhead investment in property-improving facilities has an opposing effect. Consider two sites which are identical in size and shape and indifferent in terms of location, but let one have access to a sewerage system and the other not. The former will, of course, be more in demand, its price will be higher, and hence there will be a tendency to use it more intensively than the other. In this sense, the ultimate effect of social overhead investment is analogous to that of decreasing the aggregate demand for space.

Of more direct concern to us are the policy effects which are brought about by changes in the transportation system itself, for these will alter the spatial structure of position rents and so tend to redistribute population and land values. Three general cases will be investigated: (1) the system-wide improvement of the transportation facilities of an urban region; (2) system-wide improvement in the operational characteristics of existing facilities; and (3) the effects on the total system of improvements localized within the system.

(1) Current urban transportation planning is becoming increasingly oriented toward the view that the entire urban complex is an integral system of multimodal and multipurpose transportation flows, the im-

provement of whose efficiency is an important public responsibility.
The improvement of the efficiency of the total urban system generally
involves three technical objectives: increasing the velocity of flow, increasing the capacity of the system, and mitigating congestion losses in
the system by elimination of capacity chokes. The effects on the transportation function can be summarized as follows: the general slope of
the transportation function would be diminished; the level of ingression
loss would be reduced; and the residual ingression loss would be largely
concentrated at the center.

These changes are illustrated in Figure 36. Here OCA represents
the original transportation function and ODB represents the new
transportation function resulting from the improvement of the system.
Assuming incomes and population constant over the period, the immediate effect would be vast expansion of the isotims into the hinterland
and consequent triggering of a substantial redistribution effect as summarized in Figure 37.

Here Am_1 is the original density gradient and Cm_2 is the density

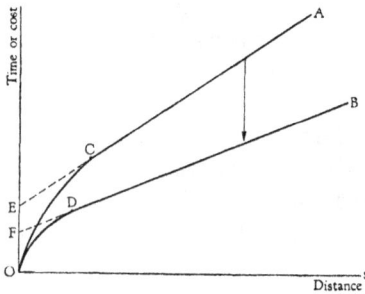

FIGURE 36.

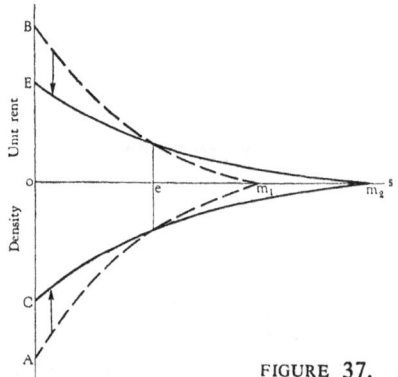

FIGURE 37.

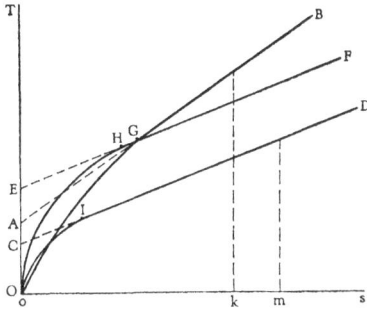

FIGURE 38.

gradient resulting from the improvement program. The effect is a re-distribution of population from the center outward with a decline in land values between o and e and an increase in land values between e and m_2.

In summary, any set of events which tends to alter the slope of the transportation function at any point will tend to alter the slopes of the density and unit rent gradients in the same direction and thus have a redistributive effect. The strength of the redistributive effect will be proportionate to the extent of the change in the transportation gradient.

(2) Improvement in the operational characteristics of a transportation system might be brought about by extensive improvement in traffic management and control and would have the direct consequence of increasing system-dominant velocities and alleviating congestion. Efforts in this direction must acknowledge that in the last analysis the capacity limitations at the core set are the basic limitations to operational improvement of the total system and that these are extremely difficult to lessen. Such improvements have two basic effects. The first effect is the direct diminution of the general slope of the function, as indicated in the change from *AB* to *EF* or *CD* in Figure 38. The second effect is ingression change which results from the capacity implications of a change in the velocity of the system. Earlier, it was pointed out that highway authorities generally accept expressions relating capacity to velocity which are generally convex upward. Typically the maximum capacity is approximately 2000 vehicles per day, the maximum velocity approximately 30 miles per hour. Accordingly, if the increase in system-dominant velocity takes place on the positively sloping region of the curve, capacity is increased and ingression diminished, as in the case represented by *OID* in Figure 38; if the change takes place on the

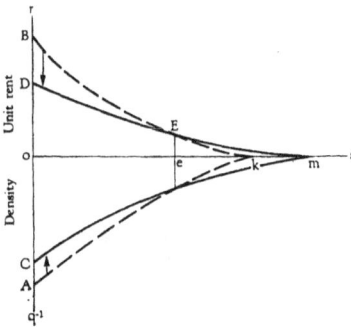

negative slope of the curve, capacity is diminished and ingression increased and a function such as *OHF* emerges. The redistributive effects of a system-dominant velocity change from *OGB* to *OID* in Figure 38 are summarized in Figure 39; the density gradient will tend to shift *Ak* toward *Cm* with a concurrent decline in land values between *o* and *e* and an increase in land values from *e* to *m*. In general, such a change will tend to facilitate a dispersal of households into the hinterlands, thus decreasing the average density of the urban area.

(3) One other case of the effects on policy of variation in the transportation function deserves comment: the general effects of localized

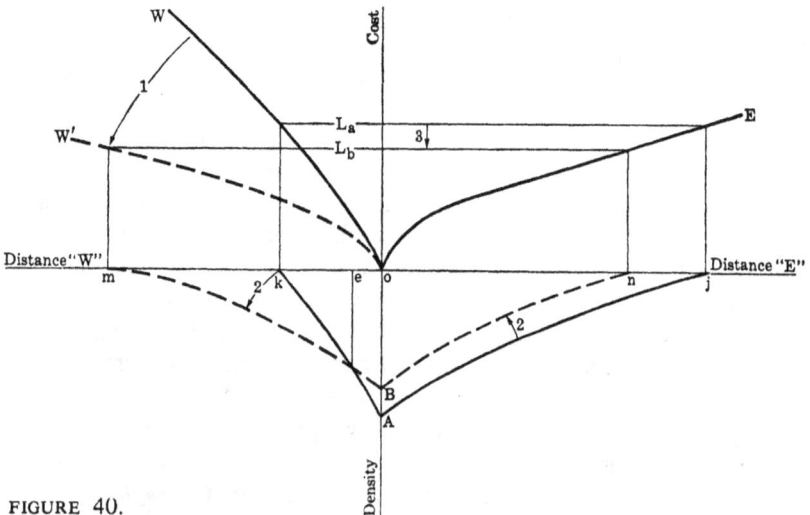

FIGURE 40.

changes in the transportation system resulting from the improvement of a single component in the system. To simplify the discussion, take a transportation system composed of two components E and W such that the E component is more efficient than the W, and suppose that the change is to increase the efficiency of W by an improvement program.

Figure 40 is a schematic working out of the consequences of the program to improve the W component. We have condensed the schematic presentation by ignoring direct unit rent effects and utilizing the horizontal or "distance E and W" axis as a common axis between the transportation function (above) and the density function (below). Conclusions regarding unit rents will tend to follow those relating to density.

Here OE and OW represent the transportation functions of the E and W components respectively before the change. Similarly Ak and Aj represent the associated density functions for a population represented by the area kAj deriving from the "pre-change" transportation functions. The pattern of effects is this:

(1) The improvement of the W component results in a new transportation function OW'.

(2) The new set of position rents[6] tends to move population (and land values) away from j toward m yielding a new population distribution mBn whose area (under constant population assumptions) is equal to kAj. Thus, densities (and values) tend to decrease to the right of e and increase to the left of e.

(3) The new distribution results in a decrease in location costs per household from L_a to L_b, and hence suggests a more efficient distribution of technology, population, and land values.

The principal usefulness of this example rests on its demonstration of how changes which are strongly localized within a transportation system may well have substantial effects throughout the entire model or in sectors quite remote from the change.

The Model as a Tool for Decision-Making

As an aid to the improvement of centralized decision-making the location model has two important contributions to make. One has been demonstrated in the preceding section—it leads to an increased ability

---

[6] As suggested in Chapter IV, competition among purchasers and sellers of land will tend to equalize location costs to all.

to anticipate some of the pertinent consequences of policy alternatives affecting urban land use or transportation. The other affords equal promise over the long run—it enables decisions to be related more sensitively to community objectives on the one hand and to important features of the policy environment on the other. It can make these contributions through its capacity for valuing the changes in the dependent variables, that is, for keeping track of the changes in the economic dimensions of the variables. It is in this sense that the location model, based as it is on the premises of economics, differs from an engineering, or technical model: where the latter can make statements about physical efficiency, only the former can say anything about how important these statements are to society and whether or not society *ought* rationally to respond to these statements.

The engineer can tell the city council that a given public investment in a highway facility can save the community a million man-hours of travel time a year, or that it can reduce the incidence of a certain type of automobile accident, but this kind of statement is not a sufficient basis for decisions. Policy-makers need further information before a rational policy can be chosen: they must know how valuable to the community are the savings in travel time or the reduced accident risk; they must know what the cost of the project is, especially the cost of capital; and they must have an awareness of alternatives for the investment of the funds. In short, whether or not this project ought to be undertaken will depend upon whether the consequences are at least as valuable as the cost of the investment and whether there are opportunities for investments whose consequences per dollar of investment are more valuable than the consequences of the one proposed. It is the relationship of the stream of benefits flowing from the project to streams of benefits foregone that is the central datum necessary for public policy decisions.

A characteristic type of system investment problem involves an extant system with a high level of congestion and a set of alternative possibilities of investment to reduce time losses in the system. The value of the ingression reduction can be construed as the social income generated by a given investment alternative. A program is justified so long as the social income that it generates exceeds the social costs of the program. Where more than one program meets this test, that program ought to be chosen whose ratio of social income, or benefits, to costs is the highest; this represents a "practical optimum" in the way of a program. Here the decision involves the substitution of capital costs for time-costs: given the cost of capital and a basis for evaluating time, we can identify this practical optimum program, which will maximize the returns enjoyed by the community as a corporate whole.

At a higher level are decisions involving long range programing for the development of an urban transportation system. Assume that a transportation system is homogeneous, that is, consists of a given number A of component movement systems whose technological characteristics are identical in that their noninstantaneous capacities are equal. The total ingression Y_A in such a system varies with the amount and distribution of demand among the components:

$$Y_A = \sum_{j=1}^{A} \frac{N_j^2}{2C_j}$$

where Y_A = the total ingression loss per cycle in the system
C_j = the capacity of component "j"
N_j = the demand on component "j" .

The total ingression among several systems with a common assembly point would be minimized if each component bore a share of the total demand in the same proportion as the proportion of its capacity to the total capacity of the several systems. A transportation system meeting these conditions is an "optimal system," since no other allocation of capacity to demand will produce a lower level of ingression cost. Thus, in a nonoptimal system a basic kind of investment decision is the "optimizing" decision addressed to reducing or eliminating the disparities in average ingression costs among the component systems. This kind of decision is similar to that discussed above; it involves the selection of an investment program with the highest social benefit-cost ratio, but with the additional restraint that it must tend to reduce the disparities, for only in an optimum system are the social costs spread equitably among all components of the system.[7]

There is another kind of investment decision associated with the comprehensive transportation system: selection of the level of average ingression around which the total system should be planned. The standard followed in transportation planning has been to match hourly demand with equal hourly capacities. This implicitly accepts an average ingression of half an hour per unit per cycle where the time distribution

[7] In a dynamic situation the operation of the urban land market may have an optimizing effect by suppressing development in high ingression sectors and encouraging it in the low ingression sectors.

of demand within the hour is highly concentrated. In other words, the matching of hourly capacities to hourly volumes will produce a low volume of ingression loss only where the peak demand upon a system is distributed throughout the hour. There is little to justify such a standard in terms of investment efficiency. If the marginal value of the road user's time-in-transit is high and the cost of capital low, a system based on such a standard could be grossly inefficient, for the social costs of the time consumed would be excessive. At the other extreme, where the value of such time is low and the cost of capital is high, such a standard might well result in the uneconomic allocation of public investment. In short, the efficient level of ingression loss in a transportation system will be a function of the costs of capital and the value of time consumed by ingression.

A host of land use issues lend themselves to this kind of partial analysis in which the transportation technology is held constant. Consider the simple case of density regulation by zoning. If the regulation successfully suppresses the development of greater household densities in areas where it is applied densities will tend to increase in those areas where effectively suppressing regulation does not exist, and thus the margin point will be pushed further into the hinterland. Location costs are increased across the board, and land values are redistributed away from the suppressed areas. The relation of short-run benefits to costs is difficult to define because of the redistribution of equities involved. The major social cost item in such a policy is the increment to location costs, the major benefit a gain in amenity (really a subsidized space preference position) by those households in the affected area, against which must be assessed the social costs of increased density in the unsuppressed areas. Whether or not the redistribution of land values is a good thing depends upon the distributive criteria of the community. This is a "rationing" effect: the rationing of one product increases the demand for, and raises the price of, all products for which it is substitutable. In this case, the rationed product is land the density of whose population is regulated, the alternative products all those lands not so regulated.

Actually, many of the typical land use issues fall within this class. The "greenbelt" approach to regulating the spatial extent of the city lends itself to this form, provided location outside the greenbelt remains possible. Here the invested city area is the restricted, superior good, while the hinterland beyond the greenbelt constitutes the unrestricted, inferior good. As the population grows the site rent component of location costs will tend to be bid up to the point where it becomes advantageous to domicile beyond the greenbelt. As a matter of fact, where urban transportation is highly developed it is unlikely that any greenbelt of practical dimensions would do other than increase land values and densities by restricting the supply of land.

In similar fashion such general alternatives as Frank Lloyd Wright's "Broadacre City" or Le Corbusier's "La Ville Radieuse" can be considered for their characteristics of economic efficiency and locational equilibrium, if the market is allowed to act as the basic mechanism of allocation. No such model will supply all the policy answers, for important objectives may lie outside its conditions. Where Wright sought a "return to the land," the economic costs of his scheme may be less important than the noneconomic results. So it may be more desirable in terms of social welfare to diminish urban densities in slum areas, in spite of consequences suggested by any model. Nevertheless, it is of crucial importance that the model consequences be known and understood if we are to avoid paying too dear a price for "noneconomic" social and cultural values.

The Model and the Future of the American City

Casting back over the discussion of the model, one is tempted to consult it, like a ouija board, and ask—what of the future? To get an answer that is even mildly satisfying, we find that we have to feed it a lot of guesses about the parameters. What is likely to happen to the marginal-value-of-leisure curves of the urban population and how are they likely to be distributed? What changes in transportation technology are imminent and what is the impact of transportation costs likely to be? What will happen to the spatial arrangement of employment? With rising incomes, what will happen to the space (density) preferences of the population?

Two features of the future labor market seem fairly certain: increasing real wages, and a contraction in the labor input per worker, both reflecting the secular increase in labor productivity. The trend toward rising incomes suggests a general upward shift of the marginal-value-of-leisure curves—the marginal hour will be more valuable in terms of the opportunity costs measured by the increased wage rate. The effect of the reduced labor input is difficult to assess, for the reduction might be brought about in any of a number of ways. At one extreme consider the later entry into (and earlier retirement from) the labor market. It is possible that part of the additional leisure time will accrue through lengthier or more frequent vacations. Finally, there is the possibility of shorter work weeks and shorter work days.

Thus there are three ways in which the marginal-value-of-leisure curve may be affected: in the upward shifting of the curve itself in response to increased labor productivity; in movement to a lower position on the curve in response to the shorter work day; and in a reduction in the annual number of work trips engaged in, responding to the reduction in the number of days worked per year. If the increase in labor productivity is taken up mainly by increased leisure, then it quite possible that the relative value of the time-costs of the journey-to-work will decline over time. If, however, most of the increased productivity is paid for in the form of income, then the annual time-costs are likely to rise in some proportion to the rise in real wages.

The technical conditions of urban transportation will face several major influences. In the first place, urban population will continue to grow rapidly. In the second place, although no major innovations can be anticipated in carrier technology in the early future, it seems likely that urban transportation systems will be dominated by the passenger automobile operating on a radial-circumferential net of freeways. To the extent that the freeway net increases over-all velocities between residences and employment centers, the time-costs of the journey-to-work will be reduced; that is to say, the slope of the transportation cost function will tend to decline. This tendency may be offset by any rise in the marginal-value-of-leisure curve, but it seems more likely that the effects of increased leisure will be in the direction of diminishing the annual costs of the journey-to-work and thus of decreasing the slope of the transportation function. If both of these trends work in the same direction, the iso-cost lines of the journey-to-work will be radically extended into the hinterland.

Finally, it would appear that some relative, if not absolute, decentralization of urban functions and employment centers will take place (although this may not be important in the near future), as technological improvement in communications and the rapid elaboration of data handling and automatic control technology weaken the customary technical linkages in the urban economy, especially those relating to the labor force. This prospect suggests a possible decline in ingression values at the center and an increase near noncentral employment centers.

These factors—decline in unit valuation of the journey-to-work, reduction in travel time of work trips, and the tendency toward deconcentration of urban economic activities—suggest a relative "flattening" of the position rent surface, the consequences of which for land values and population densities depend upon changes in space (density) preferences of the population.

It seems probable that the space preferences of households will continue to expand. The cultural value of the owner occupied, single family dwelling on its private parcel of land not only continues unabated but is amplified by the housing activities of the federal government, as well as by the land use policies of local governments. The only major check to this trend is the time it takes local governments to provide the public capital necessary for the development of public services and facilities, ranging from streets and water distribution systems to schools and parks.

The unfolding picture, then, is one of low density urban sprawl reaching out along the radial freeways far into the hinterland and nucleated by commercial and industrial development at the interchange points. At the center, densities will tend to decline—radically in the more obsolescent areas—and with declining densities go declining land values. At the very center a more highly specialized central business district appears likely, vigorous and healthy as a whole, but less dominant in the Gestalt of the urban region.

Thus, the model suggests a picture of the future of American cities not unlike that predicted by many students of urban problems. Two conclusions have a more general importance:

(1) The gradient of land values outward from the center is likely to decline sharply, with its heaviest impact on the older "gray" areas lying between the central business district and the modern suburbs.

(2) The per capita conversion of land to urban use at the periphery is likely to increase substantially, not only because of the space demand of additions to the household population but because of the "undoubling" effect of mass migration from the more central urban areas to the low density suburbs.

Urban sprawl, the subculture of suburbia, and accelerated decay around the core seem to be the major facets of the coming pattern of the large American city. If the people of our urban communities find any of these conditions—or all of them—distasteful and want them controlled, it is evident that the powerful forces now at work shaping our urban futures must be redirected and that it will take agreed-upon alternatives, sharply edged policy, and substantial development expenditures to do it.

Summary and Index of Important Expressions and Equations

---

15   62 $X = p_o(x_o) + \sum_a p_a x_a + \sum_b p_b x_b$

16   62 $X_{i-j} = \phi(s_{i-j}, n_j, p_a, x_a, p_b, x_b)$

17   65 $_k R_i + {}_k X_{i-o} = L$

18   68 $_k R_i = L - {}_k X_{i-o} = {}_k X_{m-o} - {}_k X_{i-o}$

19   74 $_k R_i = m({}_k X_{i-j})$

20   75 $_a s_{i-j} = T({}_o s_{i-j})$

21   75 $S = S({}_o s_{i-j})$

22   77 $O \leqslant \sigma \leqslant 2\pi s$

23   79 $\sigma' = \sigma'(s)$

24   81 $r_i = R_i q_i^{-1}$

25   81 $W = W' + X + R$

26   82 $R = rq$

27   85 $q_i = \left(\dfrac{\lambda}{r_i}\right)^{\frac{1}{\eta}}, \ \eta > 1$

28   85 $q_i^{-1} = \left(\dfrac{r_i}{\lambda}\right)^{\frac{1}{\eta}}$

29   85 $_k q_i^{-1} = {}_k\left(\dfrac{r}{\lambda}\right)_i^{\frac{1}{\eta}}$

30   86 $q^{-1} = \left(\dfrac{R}{\lambda}\right)^{\frac{1}{\eta-1}}$

31   86 $N = \displaystyle\int_o^m \sigma' q^{-1}\, ds$

32   87 $N = \displaystyle\int_o^m \sigma'\left[\dfrac{\phi(s_{i-o}, n)}{\lambda}\right]^{\frac{1}{\eta-1}} ds$

A Further Note on Ingression and Congestion Along Transportation Routes

The intimate relationship between ingression and congestion was illustrated in an article in the *Washington Post* of November 18, 1958:

> AEC has its own daily traffic tie-up which an employee describes as one of the most exasperating periods in each day. Up to 700 cars jockey for position on AEC's lone access road leading to Route 240. The employee said it had taken him twenty minutes the night before to get from the parking lot to Route 240. The usual time was from 12 to 15 minutes. . . .

This is both an ingression and a congestion phenomenon. The over-all capacity of a system is governed by the least capacity at any point. These capacity "chokes" give rise to a pure form of congestion loss which occurs at any point where capacity is less than in the preceding segment of the system. Here the rate of flow is diminished and units queue up waiting entry through the choke. At one extreme, management of the rate of approach of units to the choke can eliminate any congestion loss. At the other, units may approach the choke at approach capacity. In the latter case, the length of the maximum queue and the congestion loss relate to the ingression loss set by the choke capacity. Subtract from one the ratio of the choke capacity to the approach capacity; this will yield a fraction, the choke coefficient:

$$K = 1 - \frac{C_c}{C_a}$$

where C_c = capacity of the choke
C_a = approach capacity
and k = choke coefficient.

It is possible to relate both the length of the maximum queue Q and the total congestion time loss X to the system ingression loss set by the choke capacity through use of the choke coefficient:

$$X = kY$$
and
$$Q = k(N - 1)$$

where k is the choke coefficient
          Y = system ingression loss set by choke capacity
and N is the deadline demand.

In short, the *actual* objective congestion loss at a capacity choke will lie somewhere between zero and X as defined above and will vary with the rate at which units approach the choke. Congestion loss at this point will be greatest when the approach flow approximates *approach* capacity; it will be zero when the approach flow is equal to or less than the *choke* capacity. Here again, the entry decisions of the individual units will determine the scale of congestion within the limits indicated. Here ingression analysis provides directly a measure of congestion potential.

A quantitative example may serve to indicate the manner in which total ingression becomes distributed in a system. Take a path with a uniform capacity of 2000 vehicles per hour on which there is a simple deadline demand of 1000 vehicles. Assume that the full demand is asserted at the beginning of the path and is destined for the terminus of the path. Now assume (1) that there is a capacity choke in the path (e.g., a signalized intersection) which will let vehicles on the path pass only 75% of the time, and (2) that at the terminus vehicles can be evacuated from the path at a maximum rate of 1200 vehicles per hour (as though there were a second capacity choke). At the most general level this is a simple system with a total capacity of 1200 vehicles per hour, developing a total phase ingression computed as follows:

(1) $$Y = \frac{N^2}{2C}$$

(2) $$Y_c = \frac{1000^2}{2400} = 416.67 \text{ hours}$$

where the capacity for the total system is set by the least capacity of any point. If there were no chokes and the *path* capacity dominated, the ingression level would be computed thus:

(3) $$Y_c = \frac{1000^2}{4000} = 250 \text{ hours.}$$

The net result of the two chokes is to increase the total phase ingression by 166.67 hours.

Now we can determine congestion time loss X_1 at the first choke by taking the indicated ingression at the choke capacity and multiplying it by the choke coefficient k_1, where

(4) $$k_1 = \left(1 - \frac{C_1}{C_a}\right), \; C_1 \leqslant C_a$$

C_1 = choke capacity of the first choke
C_a = approach capacity to the first choke.

Then
(5) $$X_1 = k_1 Y_1$$

or $$X_1 = \frac{N^2}{2C_1}\left(1 - \frac{C_1}{C_a}\right) = \frac{N^2}{2}\left(\frac{1}{C_1} - \frac{1}{C_a}\right).$$

Where N = deadline demand,
Y_1 = indicated ingression at first choke capacity.

The *approach* capacity to the second choke is set by the *choke* capacity of the first. Hence the choke coefficient for this second choke k_2 would be defined thus:

(6) $$k_2 = \left(1 - \frac{C_2}{C_1}\right), \; C_2 \leqslant C_1.$$

Where C_2 = choke capacity of second choke,
C_1 = choke capacity of first choke,
and congestion time loss X_2 defined thus:

(7) $$X_2 = k_2 Y_2$$

$$X_1 = \frac{N^2}{2C_2}\left(1 - \frac{C_2}{C_1}\right) = \frac{N^2}{2}\left(\frac{1}{C_2} - \frac{1}{C_1}\right).$$

Then the total congestion time loss at both chokes X is defined as

(8) $$X = X_1 + X_2$$

$$X = \frac{N^2}{2}\left[\left(\frac{1}{C_1} - \frac{1}{C_a}\right) + \left(\frac{1}{C_2} - \frac{1}{C_1}\right)\right] = \frac{N^2}{2}\left(\frac{1}{C_2} - \frac{1}{C_a}\right).$$

Coming back to the example
$$N = 1000$$
$$C_a = 2000$$
$$C_1 = 1500$$
$$C_2 = 1200$$

it can now be computed that the total congestion time loss X imposed by the two chokes is

$X = 166.67$ hours or an average of 10 minutes per vehicle.

It can further be computed that this congestion loss is divided between the two choke effects:

$X_1 = X_2 = 83.33$ hours or 5 minutes per vehicle average.

This total *congestion* loss figure of 166.67 hours is identical with the additional *ingression* loss computed earlier. Thus the total loss in a system is equal to the computed ingression at minimum capacity and includes the *congestion* losses resulting from successive restrictions in capacity in the direction of movement as well as the residual ingression in the system. This formulation applies to the perfectly managed (maximum efficiency) system and is the minimum level of loss. Inefficiencies resulting from behavior of units are external to this system and would hence increase the actual level of loss experienced.

Under this simple formulation, the kind of congestion quantified is that which results from successive restrictions on capacity in the direction of movement. When capacity is uniform throughout (i.e., where $C_a = C_1 = C_2$ etc.), no congestion loss is defined and the residual ingression loss is the total loss experienced.

The Mathematics of Ingression

---

1. *System Ingression*

The quantity of ingression loss experienced by the *n-i*th unit is

$$(1) \qquad y_{n-i} = \frac{i}{C}.$$

For convenience of notation, we can modify this expression:

$$(2) \qquad y_i = \frac{n - i}{C}$$

where y_i = the ingression loss experienced by the *i*th arrival,

n = the total number of arrivals, and

C = capacity of the movement system.

If the total number of arrivals n is expressed as system demand N, the mathematical expression for total ingression loss Y in such a system can be derived as follows:

$$(3) \qquad Y = \sum_{i=1}^{N} y_i = \frac{1}{C} \sum_{i=1}^{N} (N - i) = \frac{N^2 - N}{2C}, = \frac{N(N - 1)}{2C}.$$

The average ingression loss $\bar{y}$, or ingression loss per unit demand, is then

$$(4) \qquad \bar{y} = \frac{N - 1}{2C}.$$

The marginal ingression loss in terms of demand y'_N is $\frac{dY}{dN}$, or

$$(5) \qquad y'_N = \frac{2N - 1}{2C}$$

and the marginal ingression in terms of capacity y'_c is $\frac{dY}{dC}$ or

(6)
$$y'_C = \frac{N - N^2}{2C}.$$

Where both N and C are quite large, we can approximate as follows:

(7)
$$Y = \frac{N^2}{2C}$$

$$\bar{y} = \frac{N}{2C}$$

$$y'_N = \frac{N}{C}$$

$$y'_C = -\left(\frac{N}{C}\right)^2.$$

Ingression loss being a time loss, it is expressed in units of time established by the capacity statement, which may be in terms of units per second, minute, hour, etc. Where risk factors and demand on the system are fixed, the level of ingression loss in the system can be decreased only through an increase in the capacity of the system, and this can be effected in only two ways: (1) by increasing the number of paths (and thus increasing the instantaneous capacity of the system) and (2) by modifying the technological features of the system (braking power, feedback time, mass).[1]

2. *Ingression Loss in a Simple, Interrupted System*

The problem is to define interruption ingression loss Y' in terms of the length of the system T, the position of an interruptor i, the distribution of units along the system N_i and N_T, the basic capacity of the system C, and the pulse characteristics of the interruptor p and $\bar{p}$.

(a)   Let N = the total demand in the system
(b) N_i = the first element (or interrupted) demand
(c) N_T = the second element (or uninterrupted) demand
(d) R = the pulse-interval ratio = $\dfrac{p}{\bar{p}}$,

when p = the time length of the period in which flow takes place through the interruptor, and $\bar{p}$ the period of no flow.

[1] See Appendix A.

k = the proportion of N_i moving in the saturated component; or the number of pulses of interrupted demand matched by pulses of uninterrupted demand divided by the total number of pulses of interrupted demand, thus

(e) $$R = \frac{kN_i}{N_T}, \text{ and } k = \frac{RN_T}{N_i}$$

(f) r = the ratio of interruption, or relative pulse length $= \dfrac{p}{p + \bar{p}}$

(also referred to as the interruption restraint)

(8) $$Y'_2 = \frac{(N_T + kN_i)^2}{2C}$$ (8) from the general ingression equation $Y = \dfrac{N^2}{2C}$

This is the simple ingression loss generated in the saturated sector.

(9) $$Y'' = \frac{(N_T + kN_i)}{C}(1 - k)N_i$$

(9) Y'' is the ingression imposed on the nonsaturated component by the time necessary to exhaust $(N_T + kN'_1)$ which is the demand in the saturated component.

(10) $$Y'_1 = \frac{[(1 - k)N_i]^2}{2rC}$$

(10) Y'_1 is the simple ingression loss deriving from the nonsaturated segment, from the general ingression equation (see No. 8 above). Here $(1 - k)N_i$ is the demand in the unsaturated component and r is the interruption restraint on the capacity of the unsaturated component.

(11) $$\bar{Y} = Y'_1 + Y'_2 + Y''$$

(11) Definition: $\bar{Y}$ is the total ingression loss in the system.

(12) $$\bar{Y} = \frac{(N_T + kN_i)^2}{2C} + \frac{(N_T + kN_i)(1 - k)N_i}{C} + \frac{[(1 - k)N_i]^2}{2rC}$$

(13) $N_T = N - N_i$ (13) Definition

(14) $\quad \overline{Y} = \dfrac{(N-N_i+kN)^2}{2C} + \dfrac{(N-N_i+kN_i)(1-k)N_i}{C} + \dfrac{[(1-k)N_i]^2}{2rC}$

but

(15) $\quad (N - N_i + kN_i) = [N - (1 - k)N_i]$

let

(16) $\quad (1 - k)N_i = a \qquad$ (16) Definition

then

(17) $\quad \overline{Y} = \dfrac{(N - a)^2}{2C} + \dfrac{(N - a)a}{C} + \dfrac{a^2}{2rC}$

(18) $\quad Y = \dfrac{rN^2 - 2raN + ra^2 + 2raN - 2ra^2 + a^2}{2rC} = \dfrac{a^2(1 - r) + rN^2}{2rC}$

(19) $\quad {}_iY = \overline{Y} - Y,$ and $Y = \dfrac{N^2}{2C}$

> (19) Definition: Interruption ingression loss $({}_iY)$ is the difference between the aggregate ingression loss of the interrupted system $(\overline{Y})$ and its simple ingression loss (Y).

(20) $\quad {}_iY = \dfrac{a^2(1 - r)}{2rC}$

but

(21) $\quad R = \dfrac{r}{1 - r} \qquad$ (21) from (d) and (f)

(22) $\quad {}_iY = \dfrac{a^2}{2RC}$

but

(23) $\quad a = (1 - k)N_i, k = \dfrac{RN_T}{N_i} \quad$ (23) from (16) and (f)

(24) $\quad a = N_i - RN_T$

(25) $\quad {}_iY = \dfrac{(N_i - RN_T)^2}{2RC} \qquad$ (25) Substituting (24) in (22)

where $\quad R = \dfrac{p}{\bar{p}},$

under the considerations previously posited, $o < k < 1$. Thus, interruption ingression loss varies with the location of the interruptor N_i/N_T and the pulse-interval ratio. Given a density distribution in such a system, the amount of interruption ingression depends essentially on the positioning of the interruptor, increasing as it approaches the assembly point. It depends, in the second place, on the value of R.

3. *Interruption-Ingression—Effect on Compound System*
   (Refer to Equations 8 and 9)

A realistic restraint on the value of R emerges from the intersection of two simple systems, much as in the manner of the intersection of two arterial streets, where a traffic signal regulates the alternate flows of traffic.

Let
(a) $_1Y_1$ and $_1Y_2$ = the interruption ingression loss on systems 1 and 2 respectively
(b) $_1Y_{1+2}$ = the interruption ingression loss in both systems
(c) $_1N$ = the total demand on system 1
(d) $_2N$ = the total demand on system 2
(e) $_1N_T$ and $_2N_T$ = the uninterrupted segment of demand on both systems
(f) $_1R$ and $_2R$ = the pulse-interval ratios on systems 1 and 2 respectively
(g) $_1C$ and $_2C$ = the respective capacities of the two systems
(h) q = a queueing loss constant

To find: The total interruption ingression loss in *both* systems in terms of the pulse-interval ratios, the capacities of the systems, and the distribution of demand on both systems with respect to the point of intersection.

(26) $_1Y_{1+2} = {_1Y_1} + {_1Y_2}$ (26) Definition

Let
(27) $_1N_i = {_1N} - {_1N_2}$ (27) Definition

and
$_2N_i = {_2N} - {_2N_T}$

(28) $_1Y_1 = \dfrac{(_1N - {_1N_T} - {_1R}\,{_1N_T})^2}{2\,_1C\,_1R}$ and $_1Y_2 = \dfrac{(_2N - {_2N_T} - {_2R_2N_T})^2}{2\,_2C_2R}$

          (28) See (25)

(29) $_1R = \dfrac{p}{\bar{p}} = \dfrac{_2p}{_1p + q}$ and   (29) See definition (d) in Subsection 2.

$_2R = \dfrac{p}{\bar{p}} = \dfrac{_1p}{_2p + q}$

$_1p + {_2p} + q = 1$

Then

(30) $\quad _2R = \dfrac{1 - q(_1R + 1)}{_1R + q(_1R + 1)}$

(31) $\quad _iY_{1+2} = \dfrac{[_1N - _1N_T(_1R+1)]^2}{2_1C_1R} + \dfrac{[_2N(_1R+q_1R+q) - _2N_T(_1R+1)]^2}{2_2C[1-q(_1R+1)][_1R+q(_1R+1)]}$

(31) Substituting (30) into (28) and (28) into (26)

Equation 31 answers our requirements. Given two intersecting systems, their capacities, and their distribution of demand with respect to the point of intersection and given the technological constant q, the total interruption ingression loss in the two systems is a direct function of the pulse-interval ratio R (expressed in terms of the first system $_1R$). Examination of this equation indicates that there is a case, given the demand on each system and the value of $_1R$, in which there is a value for $_1N_T$ and $_2N_T$ such that the aggregate interruption ingression loss will be zero. In all other cases, however, there is a positive interruption load loss in the system. Given all other factors except $_1R$, there is a value for $_1R$ that will minimize this interruption ingression loss. There is a different value for $_1R$ that will equate the average ingression loss in the two systems. In the special case where $\dfrac{_2N_T}{_2N} = \dfrac{_1N_T}{_1N}$ the $_1R$ which will equate the average values will also provide the minimum interruption ingression loss in the systems.

4. The Ingression Consequences of Complex Time Restraints

Assume that the arrival time restraint requires that all units be present at the destination at some time during the time period T_1-T_2. (Our time-space fixed arrival restraint hitherto has required that all units be present at the destination at the fixed time T.) So long as the demand on the system does not exceed the capacity of the system during this period, there is no ingression loss; that is to say, an arrival time can be selected for each unit, such that it will fall within T_1-T_2, and can be associated with a departure time, so that the only time loss is that of ideal travel time. Where demand exceeds capacity, however, it is necessary to satisfy the restraint that the excess demand arrive at the destination in advance of the period and there wait until the period begins. Thus ingression loss exists in this system only when demand exceeds the derived capacity of the period and can be expressed as follows:

(32) $Y = \dfrac{(N - Ct)^2}{2C}$, $N > Ct$; for all other values of N, $Y = 0$.

where $\qquad$ N is the total demand in the system,

C is the capacity of the system,

and $\qquad$ t is the elapsed time between the limits of the arrival time restraint.

An extension of this argument will govern the case in which the total demand is broken up into segments, each having a different arrival time restraint but using the same system. This is analogous to the staggering of working hours in the central business district and can be expressed as follows:

(a) $\quad$ $*$ = an operational symbol, viz. $a*b$: "a or b, whichever is larger"
(b) $\quad$ n'_a = cumulative excess demand not satisfied by capacity of periods a to x
(c) $\quad$ N_a = actual demand associated with period "a"
(d) $\quad$ C = capacity per unit of time
(e) $\quad$ n_a = excess of demand for period "a" over capacity for period "a"
(f) $\quad$ t_a = time length of period "a"
(g) $\quad$ x = total number of periods
(h) $\quad$ a = 1, 2, 3, . . . x
(i) $\quad$ m_a = the impairment of volume in period "a" resulting from cumulative excess volume in period "$a + 2$" and later
(j) $\quad$ Y_a = ingression loss developed by excess of demand over capacity in period "a"

(33) $\quad n'_a = [(N_x*Ct_x) - Ct_x] + [(N_{x-1}*Ct_{x-1}) - Ct_{x-1}] + \cdots$
$\quad\quad + [(N_{x-(a-1)}*Ct_{x-(a-1)}) - Ct_{x-(a-1)}]$.

(34) $\quad$ Let $[(N_a*Ct_a) - Ct_a] = n_a$.

(35) $\quad$ Then $n'_a = n_x + n_{x-1} + \ldots n_{x-(a-1)} = \displaystyle\sum_{b=x-(\alpha-1)}^{x} N_b$.

(36) $\quad m_a = n_x + n_{x-1} + \ldots n_{x-(a-2)} - ct_{x-(a-1)} = \displaystyle\sum_{bx-(=a-2)}^{x} n_b - ct_{x-(a-1)}$.

(37) $\quad Y_a = \dfrac{n'_a(n'_a - 1) - m_a(m_a - 1)}{2C}$

or, in high volume systems

$\quad\quad Y_a = \dfrac{n'^2_a - m^2_a}{2C}$.

(38) $\displaystyle Y = \sum_{a=1}^{x} Y_a,$

Y = total ingression loss over x periods.

Two conclusions are immediately generated from this expression:

The maximum ingression loss results when all of the t's are equal to zero, this being the case of the fixed arrival time restraint. When this is the case, the value of this expression becomes $\dfrac{N^2}{2C}$.

A minimum value for this expression exists for any set of t values such that for every pair Ct_i, N_i, Ct_i, $> N_j$.

This is in essence the case of the total ingression of N broken down into k independent systems, each having identical capacity restraints, and the value of the expression becomes

$$\frac{N^2_1 + N^2_2 + N^2_3 + \ldots + N^2_k}{2C} \; .$$

Where neither of the special conditions above exists,

$$\frac{(N_1 + N_2 + N_3 + \ldots + N_k)^2}{2C} > Y > \frac{N^2_1 + N^2_2 + N^2_3 + \ldots + N^2_k}{2C} \; .$$

Thus the expression to the right can be modified to meet the conditions specified where time-of-arrival restraints are defined as a time period, as mentioned earlier.

For Product Safety Concerns and Information please contact our EU
representative GPSR@taylorandfrancis.com
Taylor & Francis Verlag GmbH, Kaufingerstraße 24, 80331 München, Germany

9 781138 962774